Carol Chapman
Moira Sheehan
Martin Stirrup
Mark Winterbottom

www.heinemann.co.uk
✓ Free online support
✓ Useful weblinks
✓ 24 hour online ordering

01865 888058

Contents

Contents

T indicates Think about spread

Introduction

Welcome to Catalyst

This is the third of three books designed to help you learn all the science ideas you need during Key Stage 3. We hope you'll enjoy the books as well as learning a lot from them.

This book has twelve units which each cover a different topic.
The units have two types of pages:

Learn about:

Most of the double-page spreads in a unit introduce and explain new ideas about the topic. They start with a list of these so that you can see what you are going to learn about.

Think about:

Some units have a double-page spread called Think about. You will work in pairs or small groups and discuss your answers to the questions. These pages will help you understand how scientists work and how ideas about science develop.

On the pages there are these symbols:

(a) Quick questions scattered through the pages help you check your knowledge and understanding of the ideas as you go along, for example,

(a) **Use the particle model to explain why the liquid will not squash.**

Questions

The questions at the end of the spread help you check you understand all the important ideas.

For your notes:

These list the important ideas from the spread to help you learn, write notes and revise.

Do you remember?

These remind you of what you already know about the topic.

Did you know?

These tell you interesting or unusual things, such as the history of some science inventions and ideas.

After the twelve units, A–L, this book contains a lot of extra material to help you get the best results in your SATs.

First, there are six double-page booster spreads to help you fully understand the five key ideas in science (cells, interdependence, particles, forces and energy). Your teacher will suggest you work through them to improve your knowledge and move up to the next level.

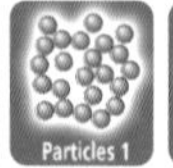

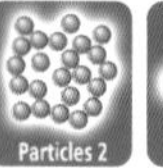

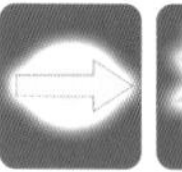

As you work through this pupil book these small symbols on some pages link to the booster spreads at the back of the book.

Next, there are five double-page revision spreads. The first one has some ideas about how best to revise for your SATs. The next four have SAT questions with notes from an examiner giving you suggestions about how to get the best possible marks.

At the back of the book:

Glossary

All the important scientific words in the text appear in bold type. They are listed with their meanings in the Glossary at the back of the book. Look there to remind yourself what they mean.

Index

There is an Index at the very back of the book, where you can find out which pages cover a particular topic.

Activities to help or check your learning:

Your teacher may give you these activities from the teacher's materials which go with the course:

Unit map

You can use this to think about what you already know about a topic. You can also use it to revise a topic before a test or exam.

Starters

When you start a lesson this is a short activity to introduce what you are going to learn about.

Activity

There are different types of activities, including investigations, that your teacher can give you to help with the topics in each spread in the pupil book.

Plenaries

At the end of a lesson your teacher may give you a short activity to summarise what you have learnt.

Homework

At the end of a lesson the teacher may give you one of the homework sheets that go with the lesson. This will help you to review and revise what you learnt in the lesson.

Pupil checklist

This is a checklist of what you should have learnt to help you with your revision.

Test yourself

You can use this quiz at the end of each unit to see what you are good at and what you might need to revise.

End of unit test Green

This helps you and your teacher check what you learnt during the unit, and measures your progress and success.

Heinemann Educational Publishers
Halley Court, Jordan Hill, Oxford OX2 8EJ
Part of Harcourt Education

Heinemann is a registered trademark of
Harcourt Education Limited

First published 2004

08 07 06 05
10 9 8 7 6 5 4

British Library Cataloguing in Publication Data is available
from the British Library on request.

ISBN 0 435 76050 5

Edited by Diona Gregory, Ruth Holmes and Sarah Ware
Designed and typeset by Ken Vail Graphic Design

Original illustrations © Harcourt Education Limited 2004

Illustrated by Graham-Cameron Illustration (Ann Biggs), SGA (Mike Lacey), Nick Hawken, Stuart Harrison, Sylvie Poggio Artists Agency (Rhiannon Powell), John Plumb.

Printed and bound in China by Everbest Printing Company Ltd.

Picture research by Pete Morris

Acknowledgements
The authors and publishers would like to thank the following for permission to use copyright material: **Question 4 p127**, taken from QCA 1997 test paper, reproduced with permission of QCA.

The publishers have made every effort to trace the copyright holders, but if they have inadvertently overlooked any, they will be pleased to make the necessary arrangements at the first opportunity.

For photograph acknowledgements, please see page vii.

Tel: 01865 888058 www.heinemann.co.uk

The author and publishers would like to thank the following for permission to use photographs:

T = top **B** = bottom **L** = left **R** = right **M** = middle

SPL = Science Photo Library

Cover: Science Photo Library.

Page 2, **T**: Sally and Richard Greenhill; 2, **BL** x2: Getty Images Ltd/PhotoDisc; 2, **BM**: SPL/Jeremy Burgess; 2, **BR**: Holt Studios; 4, **T**: Corbis; 4, **M**: Frank Lane/R Bird; 4, **B**: SPL/BSIP, Fife; 5: Harcourt Education Ltd/Peter Morris; 6, **T**: Harcourt Education Ltd/Haddon Davies; 6, **M**: Harcourt Education Ltd/Peter Morris; 6, **B**: Corbis; 7, **L**: Harcourt Education Ltd/Peter Morris; 7, **R**: SPL/Eurelios/PH Plailly; 9: Panos Pictures/Sean Sprague; 12, **T**: Corbis; 12, **B**: SPL/A Glauberman; 16, **L**: SPL; 16, **M**: John Walmsley; 16, **R**: SPL/John Daugherty; 18, **T**: Mirrorpix; 18, **B**: SPL/Quest; 20, x2: SPL; 24, **T**: Getty Images UK/PhotoDisc; 24, **B**: SPL/Adam Hart-Davies; 25: SPL/Andrew Syred; 26, **T**, **M**: Holt Studios/Nigel Cattlin; 26, **B**: Biophoto Associates; 27, x2: Harcourt Education Ltd/Gareth Boden; 28: Harcourt Education Ltd/Gareth Boden; 29, **T**: Holt Studios; 29, **ML**, **M**, **MR**, **BL**: Harcourt Education Ltd/Peter Morris; 29, **BM**: Corbis; 29, **BR**: Harcourt Education Ltd/Gareth Boden; 31: Environmental Images; 32, **TL**, **BR**: Corbis; 32, **TR**, **BL**: Getty Images UK/PhotoDisc; 34: Harcourt Education Ltd/Source unknown; 36: Corbis; 38, **TL**, **TR**: Ecoscene; 38, **B**: Papilio; 39: Oxford Scientific Films; 42, **T**: Harcourt Education Ltd/Trevor Clifford; 42, **B**: Harcourt Education Ltd/Peter Gould; 44, **T**: Alamy; 44, **B**: Harcourt Education Ltd/Peter Gould; 45: Harcourt Education Ltd/Andrew Lambert; 46, **L**: Alamy; 46, **M**, **R**: SPL; 47: Harcourt Education Ltd/Peter Morris; 48, **T**: SPL/Martin Bond; 48, **B**: Harcourt Education Ltd/Andrew Lambert; 49, **T**: Harcourt Education Ltd/Haddon Davies; 49, **B**: Chris Honeywell; 50, **T**: Harcourt Education Ltd/Andrew Lambert; 50, **BL**: Corbis; 50, **BR**: Harcourt Education Ltd/Peter Gould; 51, **L**: Harcourt Education Ltd/Peter Morris; 51, **R**: Dorling Kingsley; 52, x2: SPL/Andrew Lambert; 53: Harcourt Education Ltd/Source unknown; 55, x3: Harcourt Education Ltd/Peter Gould; 56, **T**: SPL/Jerry Mason; 56, **B**: Harcourt Education Ltd/Peter Gould; 57: Beken of Cowes; 60, x4: Corbis; 61: Getty Images Ltd/PhotoDisc; 62: SPL/John Mead; 64, **TL**, **TR**, **BR**: Corbis RF; 64, **TM**: Getty Images Ltd/PhotoDisc; 64, **BL**: Corbis; 65, **T**: Corbis; 65, **B**: SPL/Custom Medical Stock Photo/Keith; 66, **T**: Harcourt Education Ltd/Peter Gould; 66, **B**: Getty Images Ltd/PhotoDisc; 67, **T**: Getty Images Ltd/PhotoDisc; 67, **B**: SPL/Gary Parker; 68, **TL**, **TR**: Martin Stirrup; 68, **B**: SPL/Simon Fraser; 70: Getty Images Ltd/PhotoDisc; 71, **L**: Rex Features; 71, **R**: Getty Images Ltd/PhotoDisc; 74, **TL**, **BL**: SPL; 74, **TR**: Corbis; 74, **BR**: Getty Images Ltd/PhotoDisc; 75: Harcourt Education Ltd/Peter Morris; 76, **T**: Getty Images Ltd/PhotoDisc; 76, **B**: Harcourt Education Ltd/Peter Morris; 77, **T**: Ecoscene; 77, **B**: Getty Images Ltd/PhotoDisc; 78, x2: Harcourt Education Ltd/Source unknown; 81: SPL/Jean-Loup Charmet; 82, **T**: Corbis RF; 82, **B**: Sherparedside; 84: Harcourt Education Ltd/Chrissie Martin; 86, **T**: Corbis RF; 86, **M**: Corbis; 86, **B**: Bubbles/Loisjoy Thurston; 87: Last Resort Picture Library; 88, **T**: Alamy; 88, **B**: SPL/Alex Bartel; 91: NASA; 92: Corbis; 93, x2: NASA; 94, **T**: Ann Ronan; 94, **B**: SPL; 95, **T**: SPL; 95, **B**: SPL/Jeremy Burgess; 96: NASA; 97: NASA; 98: Rex Features; 100: Harcourt Education Ltd/Bea Ray; 101: Harcourt Education Ltd/Peter Gould; 102: Corbis RF; 104: SPL/Keith Kent; 105, **T**: PA Photos; 105, **M**: Rex Features; 105, **B**: Corbis; 106, **T**: Mirrorpix; 106, **B**: Alamy; 107: Honda; 108: Bruce Coleman/Pacific Stock; 110: Robert Harding; 112: Harcourt Education Ltd/Peter Gould; 113: Eye Ubiquitous/Corbis/Robert and Linda Mostyn; 115, x2: Harcourt Education Ltd/Gareth Boden; 118: Getty Images Ltd/PhotoDisc; 119, **T**: Getty Images Ltd/PhotoDisc; 119, **BL**, **BR**: Harcourt Education Ltd/Gareth Boden; 128, **T**: Harcourt Education Ltd/Andrew Lambert; 128, **BL**: Harcourt Education Ltd/Peter Gould; 128, **BM**: Getty Images UK/PhotoDisc; 128, **BR**: Harcourt Education Ltd/Roger Scruton; 129: Harcourt Education Ltd/Gareth Boden.

A1 The way we are

Learn about:
- Inheritance

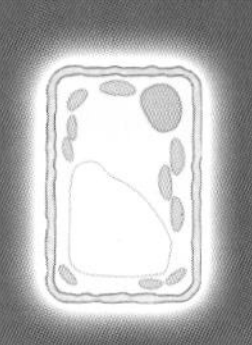

Why do we look like our parents?

If you look at the members of a family you will see that although they are all different, they have many similarities. This is because some of the same features from their parents and grandparents have been inherited or passed on to them. A baby inherits some features from its mother and some from its father. The information about these features is carried inside the nucleus of the sperm and the egg. One successful sperm fertilises the egg. The sperm nucleus and egg nucleus fuse during fertilisation to form the first cell of the new baby.

The nucleus of both the sperm and the egg contain genes. **Genes** are instructions that control the way our features develop. There are genes to control many different features. Each baby inherits half its genes from its mother and half from its father.

Every sperm and every egg contains a different set of genes. When a sperm and egg fuse to form a fetus, there is a completely new combination of genes. That is why the children of two parents are not the same, except for identical twins, although they may have many similarities. Each baby is unique: it resembles its parents in some ways, but is not exactly like either of them.

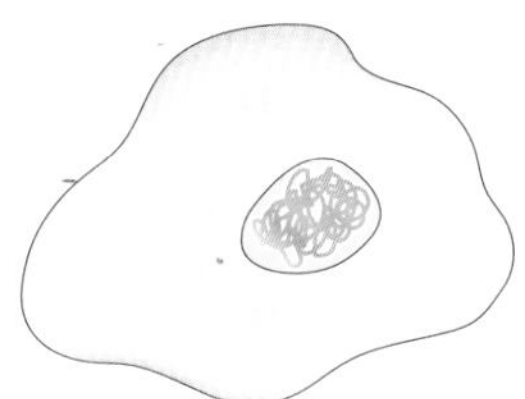

The nucleus of the cell contains thousands of genes.

a **Why do you think a baby does not look half like its father and half like its mother?**

Only humans?

Features are inherited in other species too. Look at the photos on the right.

b **(i) Describe one feature that has been inherited by the rats.**
(ii) Describe one feature that has been inherited by the pea plants.

In plants, the nucleus of the pollen grain and the nucleus of the egg cell fuse during fertilisation. The fertilised egg cell will grow into a new plant and its features will be controlled by the genes inside the pollen grain and the egg cell.

Twins

In humans, when a sperm fertilises an egg, the fertilised egg may split into two embryos. Identical twins are formed. Both twins have come from the same sperm and egg, so they have the same genes and the same features.

If there are two different eggs at the same time, and each is fertilised by a different sperm, the twins will have different genes and different features. They are non-identical twins.

c **Can identical twins be of different sexes? Explain your answer.**

Inherited variation

There are genes to control all of your features. Your genes control, for example, whether you will have:

- curly or straight hair
- blue or brown eyes.

Some of these features controlled by your genes are either one thing or another – there is no 'in-between'. Look at bar chart **A**. For example, some people have ear lobes that are attached to the side of their face, and other people have unattached ear lobes. Which type have you got?

Other features show a range. Look at bar chart **B**. For example, for height there is a range from very short to very tall with lots of heights in between. If a baby inherits genes for tallness from both parents, it will probably grow up to be tall. If it inherits genes for tallness and genes for shortness, we can't be sure what effect these genes will have on the person's height.

A

Number of pupils

Attached ear lobes

Unattached ear lobes

Environmental variation

Identical twins have exactly the same genes, but they sometimes look different when they grow older. Studying identical twins gives us a lot of information about how features are affected by the environment.

For example, a pair of identical twins may both inherit genes for tallness, but one may have a poorer diet than the other. The one with the healthier diet will grow taller. If one twin does not go to school, he or she may not reach the same level of intelligence as the twin who is educated.

The environment can affect plants too. Bonsai trees are miniature trees. They have the normal genes for tallness, but are small because they have been grown in small pots.

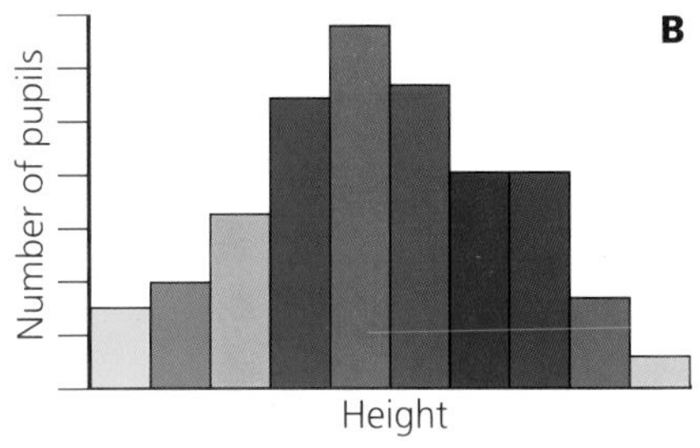

All these examples show how genes and the environment can cause variation in features. This means some features are a result of both inherited variation and environmental variation.

Family trees

The diagram on the right shows the family tree for the Foster family. The children inherit their hair colour from their parents.

Diane Dave Mabel Jim

Laura Rose Julie Mark Pat

Jenny Leigh Hannah

d **Examine the family tree. From which people has Leigh inherited genes for red hair?**

e **Laura and Rose are identical twins. Why have they both got dark hair?**

f **Suggest two other features that will be the same for Laura and Rose.**

g **Laura weighs 65 kg and Rose weighs 50 kg. Suggest why their weights are not identical.**

Questions

1. Explain how we inherit our features from our parents.
2. Where in the cell are the genes?
3. Here is a description of a Year 9 girl: 'blue eyes, not very tall, olive skin, red hair, intelligent, attached ear lobes'. Which of these features may have been affected by her environment?
4. Draw diagrams to explain how identical twins and non-identical twins are formed.
5. Scientists often study identical and non-identical twins as they grow up. Suggest some of the ideas they might investigate.
6. Two seeds from the same apple were planted in different areas. One tree had large apples, and the other tree had small apples. Describe two factors that might have caused this variation.

For your notes:

- An organism inherits features from both its parents. These features are controlled by **genes**.
- Identical twins have the same genes because they come from the same sperm and egg.
- Non-identical twins are formed when two sperm fertilise two eggs.
- Both inheritance and the environment cause variation between the members of a species.

A2 Choose your parents

Learn about:

- Selective breeding in animals

Select a winner

The racehorses in the photo are all members of the same species. Some are taller than others and they have different coloured coats. But racehorse breeders are mostly interested in how fast they can run. They select the fastest runners to breed the next generation of winners.

Do you remember?

Members of the same species have similar features. They can reproduce together to continue the species.

Another pint of milk

Like other mammals, the cow produces milk to feed her calves after they are born. This is called **lactation**. The milk gives young mammals a continuous supply of nutrients, and also protects them from disease. Young animals do not have to go out and find food, so they are protected from predators. The amount of milk a cow produces is called the **milk yield**. Farmers sell milk for a profit, so high milk yield is a **desirable feature** in a cow. This feature can be passed onto the next generation.

High milk yield is a feature that is inherited from both the bull and the cow. Farmers take a bull that has genes for high milk yield and mate it with the cows that produce the most milk. They may produce new cattle with higher milk yields. The farmers are selecting the bull and cows that have the desirable features they want to pass on, and breeding from them. This is called **selective breeding**.

Selective breeding

Instead of mating a bull and a cow at the farm, scientists at a breeding station may choose a bull that can pass on high milk yield and sell its semen to farmers. The farmer puts the semen into the cow's vagina through a long tube. We call this **artificial insemination**.

This table shows the statistics for cows bred from Gemidge and Goldfinger, two different bulls.

Feature	Yield compared with the average	
	Bred from Gemidge	Bred from Goldfinger
milk	+902 kg	+1203 kg
fat	+23.3 kg	+30.0 kg
protein	+29.1 kg	+31.8 kg
milking speed	average	fast
temperament	very good	good

a **Which bull do you think passed on genes for the highest milk yield, Gemidge or Goldfinger?**

b **Milk is used to make products such as cheese and yoghurt. Cheese is made from milk protein. Which bull's semen would you use to produce a cow if you wanted to use its milk to:**
(i) make more cheese? (ii) make low-fat yoghurt?

c **(i) Which bull's daughters have the better temperament?**
(ii) What advantage do you think it is to the farmer to have a cow with a good temperament?

Other features

If you select two parents that each have different desirable features, such as resistance to disease or strong legs, you can try and produce new varieties of cattle with all of these features. The wild ancestors of our domestic cattle were the aurochs. They are now extinct, but there are cave paintings of them. Longhorn cattle are very similar to the aurochs. They were once popular as dairy and beef cattle because they could walk a long way to market. Longhorns are now a rare breed, and it is important that we protect them from becoming **extinct**. When a species becomes extinct, it dies out altogether and we lose its useful genes.

Did you know?

To collect the bull's semen, the bull may be introduced to a frame that is made to look like a cow, with a cow's hide over it and a rubber vagina inside. The bull tries to have sexual intercourse with it, and the semen is collected in the artificial vagina. The semen is frozen in straws until it is needed. Every second a calf is delivered at Avoncroft breeding station that was fathered by a Dutch bull called Sonny Boy, but he has never met a cow!

d **(i) Why do you think longhorn cattle are less popular as dairy and beef cattle now?**
(ii) Why do you think it is important to protect them?

Questions

1 Explain what we mean by 'desirable features'. Give an example.

2 List the steps a farmer can take to selectively breed cattle for high milk yield.

3 Make a list of features in cattle, other than milk yield, that a farmer might selectively breed for.

4 Goldfish are found in lots of different colours that look attractive in aquariums. Suggest how the more unusual colours might be introduced by selective breeding.

5 Imagine you are a vet. Write a letter to Farmer Brown giving advice on how he can improve his herd's milk yield. He has cows that give a creamy milk but needs to supply more customers.

For your notes:

- **Desirable features** are features you want to pass on.
- We can select parents with desirable features to produce new varieties of animals with these same desirable features. This is called **selective breeding**.

A3 Choice vegetables

Learn about:

- Selective breeding in plants

Sweet and firm?

Think about all the different varieties of tomato that you can buy in a supermarket – different sizes, colours, textures and flavours. Plant breeders have selected different features to produce all of these varieties.

A loaf of bread

Wheat is grown to make flour. Wild wheat was one of the earliest plants cultivated by humans. Scientists think that wild wheat was the offspring of wild grasses. New varieties of wheat have been produced by selective breeding that give a high yield of good-quality grain. Modern wheat varieties are also resistant to disease.

The wheat flower is normally **self-pollinated**. The nucleus of the pollen cell from the male part of the flower fertilises the nucleus of the egg cell in the same flower. When new varieties are produced:

- The parents are selected for desirable features.
- The parent plants are planted in two rows, one chosen to be the females and one chosen to be the males.
- A chemical is sprayed onto the female parent plants to sterilise the male part of the flower. The male plants are not treated.
- The pollen is carried by the wind from the male parent plants to the female parent plants.
- The offspring produced will have desirable features from both parents.

Do you remember?

Pollination happens when pollen grains are transferred from an anther to the stigma in a flower.

Research

Moira works in a plant breeding station. She is trying to develop new and better varieties of broccoli to sell to supermarkets.

She has identified these desirable features: bright green colour, large flower heads, frost resistant, nice taste and long shelf-life.

a (i) **Suggest a reason why she has chosen each one of these features.**
(ii) **Which feature do you think is the most important to breed selectively for?**
(iii) **What do you think Moira should do to breed selectively for this feature?**

Feed the world

Scientists have used selective breeding to develop varieties of food crop that can grow in dry conditions and in poor soil. There is a variety of carrot that can grow in very stony ground. This kind of research has helped to provide food in countries where droughts are common.

Did you know?

The Inca Indians in Peru were the first people to cultivate potatoes for food, as early as 200 BC. These potatoes looked very different from modern potatoes. They were all different colours – imagine red chips!

Genetic engineering

Scientists can now take genes out of one species and put them into another to give them desirable features. This is called **genetic engineering**. It is a modern form of selective breeding. Scientists have put genes into soya beans so that the plants can make chemicals which kill pests trying to feed on them. They have put human genes into sheep cells so that the sheep produce milk with a human protein in it. This can be used to treat people with the lung disease emphysema.

Did you know?

In **genetically modified (GM) food**, genes may be turned off or new genes may be put in to change the features of the plant. In genetically modified tomatoes, the gene that makes them go soft has been turned off so they stay firm for longer.

Cloning

Clones are genetically identical organisms – all their genes come from one parent. Taking cuttings is one method of cloning plants. It has been used for many years. A section of a plant is cut from the parent plant and used to grow a new plant. Only one parent is involved. Reproduction without the nuclei of sex cells fusing together is called **asexual reproduction**.

In 1996, a team of scientists lead by Dr Ian Wilmut took a cell from the mammary gland of an adult sheep and removed the nucleus. They removed the nucleus from a sheep egg cell and replaced it with the mammary gland cell nucleus. The egg grew into an embryo which was put into the uterus of another sheep, which gave birth to Dolly (below). Dolly was the first mammal that was cloned from an adult cell.

Did you know?

Identical twins have identical genes!

Questions

1. Draw a diagram to show how you can selectively breed wheat.
2. Explain why selective breeding is as important in plants as it is in animals.
3. When breeding wheat, how does the farmer make sure that some of the wheat flowers are pollinated by pollen from the selected plants?
4. 'Genetic engineering might replace selective breeding.' Suggest a reason for this statement.
5. Scientists might soon be able to clone human beings. Do you think this would be a good idea? What problems would it cause?

For your notes:

- Desirable features in plants can be passed on by selective breeding to produce new varieties of plants.
- **Clones** are genetically identical organisms.

A4 Happy families

Think about:
- Probability

Boy or girl?

All the information that decides whether a baby is male or female is carried inside the sperm and the egg. The sperm can have a male Y factor or a female X factor. The egg can only have an X factor.

If an X factor egg is joined by an X factor sperm to give XX, the baby will be a girl. If an X factor egg is joined by a Y factor sperm to give XY, the baby will be a boy.

X

a **If a couple's first child is a boy, do you think their second child is more likely to be a girl?**

Heads or tails?

If you spin a coin, there are two possible outcomes: heads or tails. Each time you spin a coin there is an equal chance of getting heads or tails. If you get two heads in a row, it is called a run of two. Three heads in a row is called a run of three, and so on.

b **Work with a partner. Try spinning a coin 25 times. Write down whether you get heads or tails each time. Count up how many runs of two, three and four you got.**

For every baby that is born, there is an equal chance or **probability** of its being a boy or a girl. The possible combinations of X and Y factors are XX and XY. There are two possibilities, just like spinning a coin. If you have five girls in a row, it is like a run of five heads. The chance of having a girl is always one in two, regardless of how many girls you already have.

Any choice?

The torn and faded pages of a Chinese manuscript tell the story of Chan, a rich emperor who desperately wants a son. He is very old fashioned and believes that a son must inherit his wealth. Chan and his wife Jade have five daughters. Chan is hoping for a sixth child – a son. Jade thinks that their family is large enough, and anyway a sixth child might be another girl.

One day, Jade thinks of an intriguing plan to change her husband's mind. She makes three types of card.

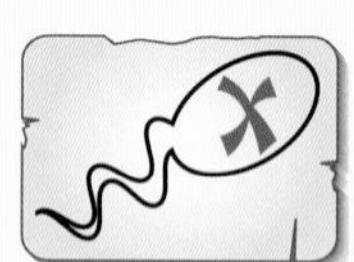

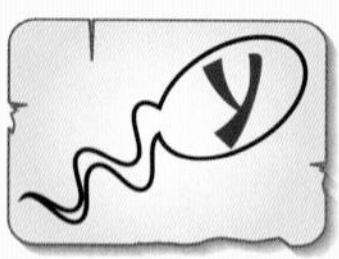

'My dear husband, only you have the power to decide whether our next child will be a boy. Each of your sperm has either the Y factor for a boy or the X factor for a girl. I can only make the X factor.' Chan is flattered that he should have such power and listens carefully.

'There are 10 male cards. 5 have the Y factor and 5 have the X factor. There are 10 female cards. They all have the X factor. I will shuffle the cards. You must choose one male card and one female card each time.

'If you choose:

it is the same as having a female child.

'If you choose:

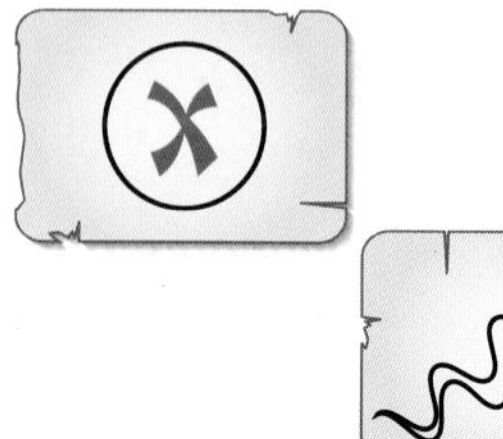

it is the same as having a male child.'

The emperor makes his first choice:

c **What is the sex of this 'child'?**

The emperor puts back the cards. He makes four more choices in this way, and each time the combination is female. 'Fascinating! This is just like our family!' Chan exclaims in surprise.

d **Do you think this result is surprising? Explain your answer.**

The bargain

'Now you must make your sixth choice,' said Jade. 'If your sixth choice is male, I will agree to having another baby, but if your sixth choice is female, our family is complete and all our wealth will be divided between our five daughters.'

e **What do you think the probability is of the sixth choice being male?**

f **Do you think Jade's card game was a good idea? Explain your reasons.**

g **How many children do you think the emperor and his wife would need to have to be sure they had a boy? Explain how you could show this with the cards.**

Family planning in China

Shanghai, China's largest city, has more than 2500 people living in each square kilometre. An estimated 16 million babies were born in 2003 in China, compared with only 4 million in the USA. Because the growing population has caused problems, the country has a family planning policy. Late marriage and late childbirth are advocated and couples normally have only one child. The family planning programme started in the early 1970s and it is estimated that China would have 260 million more people than it has now if the programme had not been adopted.

h **Some Chinese people have argued that the limit should be one son, rather than one child, as daughters are not popular there. Others say that more than 98% of the families trying to have one son would have up to six children. What do you think?**

Questions

Discuss the questions with a partner. Write down the answers.

1 **a** If a sample of a man's semen contains 300 million sperm, how many of the sperm do you think will have: **(i)** the Y factor? **(ii)** the X factor?

b A woman produces one egg each month. How many of them have the X factor?

2 **a** Estimate the number of eggs a woman produces in her lifetime.

b Why do you think that a man produces so many more sperm than the number of eggs a woman produces?

3 The Smith family have three children. The oldest one is a boy and the two younger girls are twins. What do you think the probability is of their fourth child being a boy?

4 Explain why the average family does not have equal numbers of boys and girls.

5 A London clinic claims to have developed a new technique for separating out sperm carrying the Y factor. What do you think of this new development?

B1 Are you fit?

Learn about:

- Fitness

What is fitness?

Fitness means different things to different people. One measure of fitness is how well your body supplies your cells with glucose and oxygen for respiration, and how well it gets rid of the carbon dioxide and water produced by respiration. The symbol equation for respiration is:

$6O_2 + C_6H_{12}O_6 \rightarrow 6CO_2 + 6H_2O$ energy is released

Do you remember?

Respiration happens when glucose and oxygen react in your cells to release energy.

oxygen + glucose → carbon dioxide + water energy is released

Supplying your cells

When you exercise, your muscle cells respire more to release more energy. To do this, they need more oxygen and glucose, and they need to get rid of more carbon dioxide and water. Your respiratory system, your circulatory system and your digestive system work together to increase the supply of oxygen and glucose.

- Your digestive system breaks down the nutrients in your food into small molecules like glucose. Glucose is absorbed into the blood and transported to your cells for respiration.
- Oxygen enters your body and carbon dioxide leaves your body through your respiratory system. If you breathe faster, you take in oxygen and get rid of carbon dioxide more quickly.
- Blood carries substances including oxygen and glucose around your body in your circulatory system. **Red blood cells** are specialised to carry oxygen for respiration. For example, they are very flexible so they pass easily through small blood vessels.

If your heart beats faster, it pumps the blood to and from your cells more quickly. This means your cells can respire more quickly. You can measure how fast your heart is beating by taking your pulse.

Do you remember?

A balanced diet contains the correct amounts of nutrients (carbohydrate, fat, protein, vitamins and minerals), along with fibre and water, to keep you healthy. Deficiencies of nutrients make people ill.

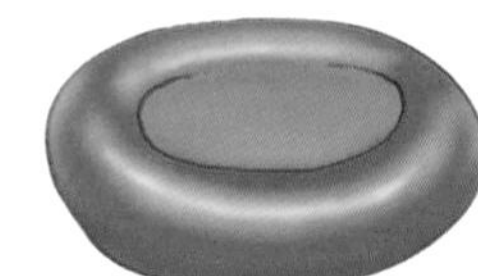

a **The graph shows the breathing rate of a man throughout a 24-hour period. Why do you think his breathing rate is low between midnight and 6 am?**

b **(i) The man is a gardener. Why does his breathing rate go up when he gets to work?**

(ii) Why does it go down in the middle of the day?

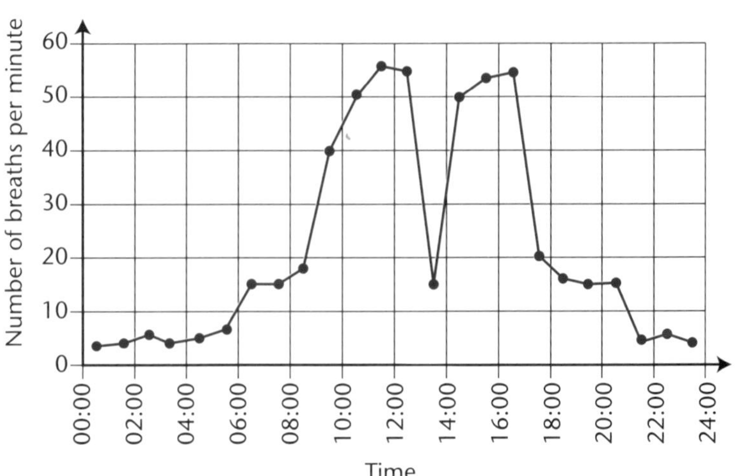

Why are some people so fit?

This graph shows James's and Aisha's heart rate before, during and after finishing a cross-country run.

Aisha exercises regularly. This means her heart and lungs have got stronger, and work more efficiently.

- Each time Aisha breathes, she takes in more oxygen and gets rid of more carbon dioxide than James.
- Each time Aisha's heart beats, it pumps more blood than James's heart.

Aisha's breathing and heartbeat are not as fast as James's but they still supply enough blood to her muscles during exercise. Aisha is fitter than James.

You can recognise whether someone is fit because:

- Their resting pulse rate is low.
- During exercise, their pulse rate does not increase so much.
- After exercise, their pulse rate quickly returns to normal.

C Which of these ideas do you think is the best way to measure fitness? Explain why.
A Measure your pulse rate at rest.
B Measure your breathing rate at rest.
C Find out how long it takes for your pulse rate to go back to normal after exercise.
D Measure how hard you can push against some bathroom scales.
E Measure the size of your upper arm muscles.

Getting fit

To get fit, your circulatory system, breathing system, digestive system, muscles and bones must be working properly. This means eating a balanced diet, avoiding smoking, drugs and alcohol, and getting regular exercise. Health experts recommend you exercise at least three times each week to be fit.

Different types of exercise can improve your fitness in different ways. Some, such as swimming, are good at improving your strength. Others, such as gymnastics, are good at improving suppleness. Jogging can improve your stamina.

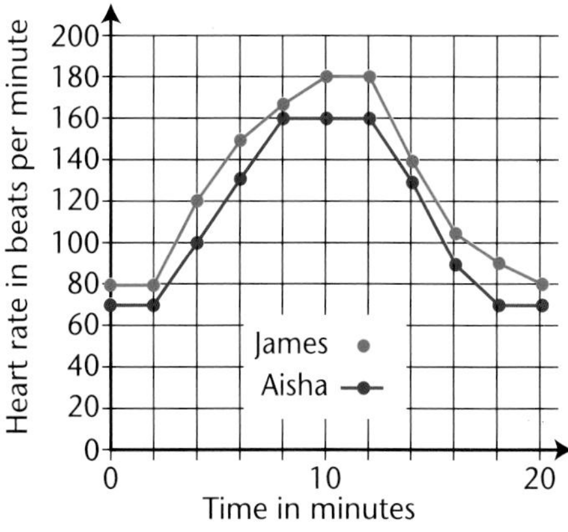

Suppleness means being able to move your body and limbs easily.

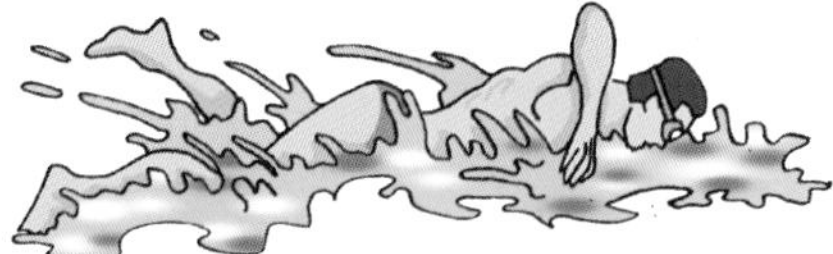

Strength is the ability of your muscles to exert a force.

Stamina is your ability to keep going when you exercise.

Questions

1. **a** What do your cells need in order to release energy from food?
   b What else do your cells produce when they release energy from food?
2. Explain why your heart beats faster during exercise.
3. Your friend has entered for the London marathon. Explain to her why it is important to train, and how training will improve her fitness.
4. Imagine you are an oxygen molecule travelling around your body. Begin by entering the mouth, and tell the story of your journey until you reach a muscle cell and get used in respiration. Try to include as much detail as possible.
5. Design a weekly fitness programme for a busy Year 9 pupil to do without needing any special equipment.

For your notes:

- **Fit** people can get enough oxygen and glucose to their cells for respiration, even when they are exercising.
- All your organ systems, particularly your digestive system, respiratory system, circulatory system and muscles and bones, must work properly for you to stay fit. Your lifestyle can affect how well these organ systems work.
- Different types of exercise improve your fitness in different ways.

B2 Breathing and smoking

Learn about:

- Breathing
- Smoking and health

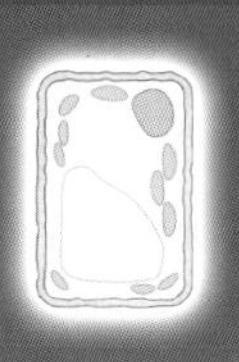

How do we breathe?

You breathe to take oxygen into your body for respiration, and to get rid of carbon dioxide produced by respiration. Put your hands on your chest, and feel your ribcage. It is wrapped around your lungs. Take a deep breath in and out. Feel how the ribcage moves. Look at the diagrams below to see what happens.

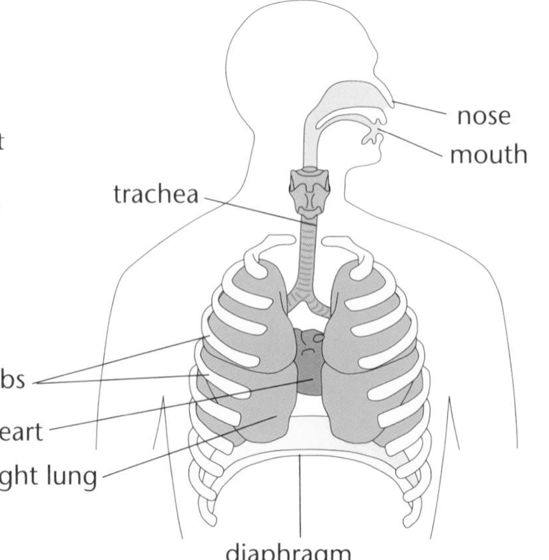

Do you remember?

Your respiratory system contains millions of tiny alveoli. In these alveoli, oxygen enters your blood and carbon dioxide leaves it.

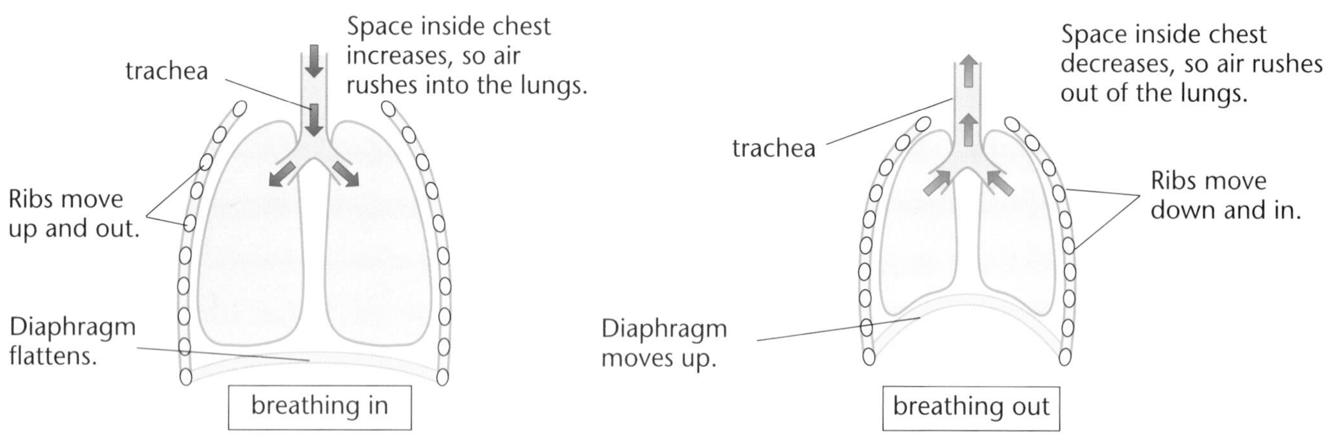

Life support

People who are seriously ill may be kept alive using a life-support machine. This helps to keep their heart and lungs working properly, to make sure all the cells in their body are supplied with oxygen.

The actor Christopher Reeve starred in the *Superman* movies in the 1980s. He had a horseriding accident that left him paralysed from the neck down. This means he cannot breathe on his own. Instead, his life-support machine pushes air into his lungs and forces his ribs up and out. When it stops pushing, his ribs move back down and in, and he breathes out. Because a machine breathes for him, his speech is very stilted. He can only say something when air leaves his lungs.

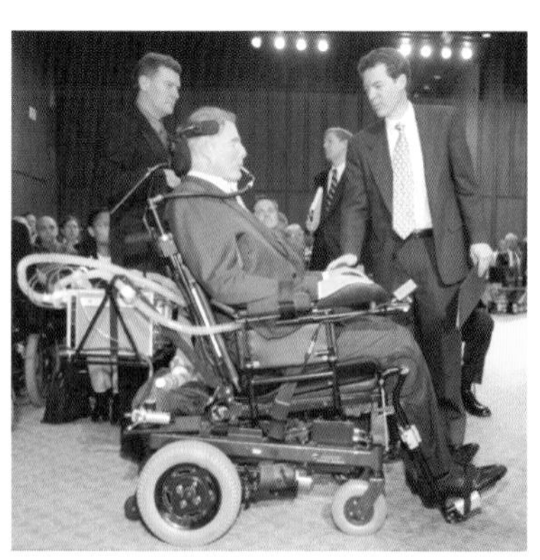

Understanding a cigarette packet

If you smoke, you breathe in a lot of harmful chemicals. These damage the alveoli in your lungs. Look at this photo of lungs from a healthy person (far left) and from a smoker (left).

The damage caused by smoking reduces the oxygen supply to your cells. The cells can't respire as much, and release less energy, so you feel tired. Many smokers breathe quickly to try to get enough oxygen. They get out of breath quickly.

Cigarette smoke causes illnesses that can seriously damage your health. Smokers are warned about the effects of smoking on cigarette packets.

Stopping smoking reduces the risk of fatal heart and lung diseases

Tar causes lung cancer and bronchitis. It damages the alveoli, so it's harder for oxygen to get into the blood. Look at the picture. You can see the unhealthy alveoli have a smaller surface area, stopping them working properly.

Poisonous carbon monoxide reduces the amount of oxygen your red blood cells can carry. ***Haemoglobin*** *is the chemical in red blood cells that carries the oxygen. Carbon monoxide stops haemoglobin carrying so much oxygen.*

Smoking causes fatal lung cancer

Lung cancer is a ball of cells that uses up space in your lungs. This leaves less space for alveoli, and less gas exchange happens.

Smoking can cause a slow and painful death

Dirt and microbes in the air you breathe in stick to sticky mucus on the inside wall of your trachea and bronchi. Specialised cells called ciliated epithelial cells use their cilia (tiny hairs) to sweep the mucus up and out of the respiratory system. If you smoke, the cilia are destroyed by tar, and the mucus, dirt and bacteria build up.

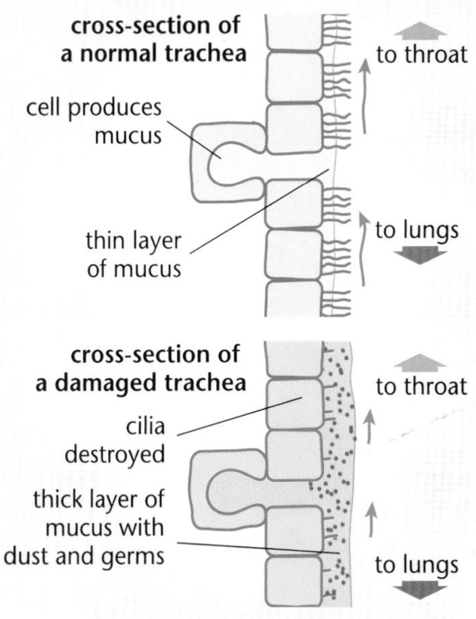

Smoking when pregnant harms your baby

A mother's blood carries oxygen to her unborn baby. Smoking reduces the oxygen supply to the baby's cells. This stops the baby developing properly.

Smoking is highly addictive

Nicotine makes it hard to give up smoking. Nicotine contributes to blood vessels narrowing, increasing your blood pressure and giving you heart disease.

Smoking clogs the arteries and causes heart attacks and strokes

Air pollution

In big cities the air may be polluted by carbon monoxide and carbon particles in smoke. These can cause asthma. People who work in some factories or mines can suffer breathing difficulties or lung diseases such as emphysema because of the dust or other substances in the air.

Questions

1 **a** Write down the parts of the respiratory system that carbon dioxide travels through on its way from the alveoli to the mouth.

b Describe how the ribs and diaphragm move to help push air containing carbon dioxide out of the lungs.

c Explain how carbon monoxide affects the ability of blood to carry oxygen.

2 Explain how: **a** nicotine **b** tar can affect your health.

3 Write a script for your favourite TV soap where two characters are arguing about smoking. Use as much scientific information as you can to explain why it is bad for your health.

For your notes:

- When you breathe in, the space in your chest increases and air rushes in. When you breathe out, the space in your chest decreases and air rushes out.
- Cigarette smoke contains chemicals that reduce the amount of oxygen reaching your cells, and can seriously damage your health.

B3 Drugs and alcohol

Learn about:

- Drugs
- Alcohol

What is a drug?

These people like taking drugs. A **drug** is any substance that changes how your body works, or alters the way you think and feel.

- Every day Edward takes caffeine, Jill takes alcohol and Hassan takes nicotine. These drugs are legal and are called **recreational drugs**.
- Joanna is taking antibiotics for a week. These are also legal and are called **medical drugs**. A doctor or pharmacist gave them to her because she is ill.
- Some people also take **illegal drugs**. They harm the body and may cause death.

a **It is illegal to use medical drugs without a doctor's permission. Why do you think this is?**

All drugs have **side effects**. These are not the main effects of the drug, like curing your illness. They are other effects the drug has, such as making you feel sleepy or sick. Side effects may damage your organs as well. If you take too many drugs, the side effects can make you very ill or even kill you.

Many drugs are **addictive**. When a drug's effects have worn off, you may feel bad and need more of it to feel good again. The bad feelings are called **withdrawal symptoms**. Caffeine, nicotine and alcohol make people feel they need a coffee, a cigarette or a glass of beer or wine. Some medical and illegal drugs can be just as addictive.

Finding out about drugs

A drugs worker came to Amy's school. He said there are three main types of drug.

- **Depressants** slow down the brain, make you react less quickly, and make you feel relaxed and sleepy.
- **Stimulants** do the opposite. They speed up the brain, make you react more quickly and make you more alert.
- **Hallucinogens** make you see, feel or hear things that don't really exist.

Example	Why people take it	Side effects
ecstasy (illegal)	Makes them feel energetic and happy.	Unable to sleep. Lack of appetite. Overheating while dancing can cause death.
heroin (illegal)	Gives a warm, relaxed and happy feeling.	Nausea, vomiting. AIDs from infected needles. Can lead to unconsciousness, coma or even death.
cocaine (illegal) amphetamines (illegal)	Make them feel confident and cheerful, and as if they have a lot of energy.	May cause depression and extreme nervousness. May cause heart failure. Sniffing cocaine can damage the lining of the nose and make it rot away.
LSD (illegal)	Makes them see, feel or hear things that aren't really there.	Loss of appetite, sleeplessness, tremors. May cause feelings of terror and panic.
nicotine (legal)	Makes them feel happy, alert and contented.	Increases the heart rate and blood pressure, leading to heart disease.
cannabis (illegal)	Makes them relaxed and slows down reactions.	May lead to confusion and disorientation. May cause lung cancer and bronchitis.

b **Write three lists of drugs from the table above, using the headings depressants, stimulants and hallucinogens.**

c **Which have the more serious side effects, legal drugs or illegal drugs?**

Alcohol

Alcohol is a depressant. After drinking alcohol, people feel less shy and more relaxed. Unfortunately, alcohol has other short-term side effects on the body. You can see these in the table.

Alcohol is an addictive drug, so people can become addicted to it. These people are called alcoholics, and often drink a lot regularly. This can lead to long-term side effects. A pregnant woman should not drink alcohol as it could harm her baby.

Alcohol is removed from the blood by the liver. The liver can remove about one unit of alcohol per hour. If a person has drunk 6 units of alcohol, it will take about 6 hours before the blood returns to normal.

If the liver has to process too much alcohol over a number of years, it becomes damaged and less efficient. It may even stop working altogether.

d **How long would it take the liver to remove the following amounts of alcohol?**
(i) 3 units
(ii) 2 pints of lager
(iii) 5 glasses of wine

Short-term side effects	Long-term side effects
dehydration (from producing a lot of urine)	hepatitis (swelling of the liver)
heat loss	cirrhosis of the liver (damage to the liver)
slower reactions	cancer of the liver, mouth, throat and oesophagus
may cause blurred eyesight and slurred speech	brain damage
may cause loss of balance	high blood pressure
	stomach ulcers
	depression
	sexual difficulties (impotence)

Did you know?

Alcohol is measured in **units of alcohol**. All the drinks in this picture contain one unit of alcohol. It is illegal to drive after drinking more than two or three units, depending on your size.

$\frac{1}{2}$ pint of beer or lager

or

single measure of spirits

or

small glass of wine

= 1 unit of alcohol

Questions

1 You can classify drugs in several different ways. For each of the following groups, make a list of all the drugs you have read about in the group. Add some of your own if you can:
a medical drugs **b** recreational drugs **c** illegal drugs.

2 Athletes are tested to check they have not used drugs to improve their performance.
a Some female athletes take drugs to make their muscles bigger. One side effect of these drugs is hair growth on the face. Explain the term 'side effects'.
b Athletes who take drugs may become addicted to them. Explain what this means, and why it can be difficult to stop taking the drug.

3 Write an information leaflet for Year 6 pupils about the effects of alcohol on the body.

For your notes:

- A **drug** is a substance that changes how your body works, or alters the way you think and feel.
- Many drugs are **addictive** and have different short-term and long-term **side effects** on your body. When you stop taking addictive drugs you suffer **withdrawal symptoms**.

B4 Injury time

Learn about:
- Bones and joints
- Injuries to bones and joints

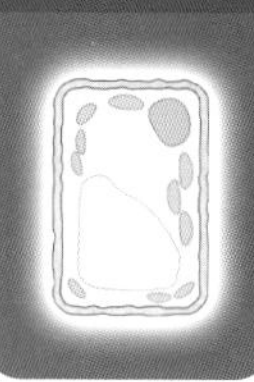

Broken bones

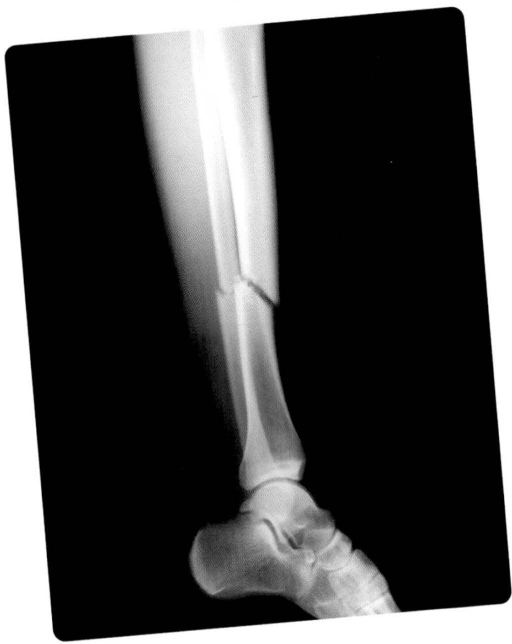

Anna broke her leg while playing football. She was carried off the pitch, and rushed to hospital. The doctor set her leg in plaster so the broken pieces can grow back together. You can see her X-ray above left.

The **skeleton** has many bones. Some of them keep Anna upright and allow her to move around. But a broken leg can't support Anna, so a plastercast gives support until the bone mends. Some parts of the skeleton protect delicate organs.

(a) Look back at page 12. Which part of the skeleton is protecting the lungs?

Bone is made mainly of calcium salts. The calcium makes the bones strong and hard. That's why you should eat calcium as part of a balanced diet. In the bone are living cells, so bones have arteries and veins to carry blood to and from the cells.

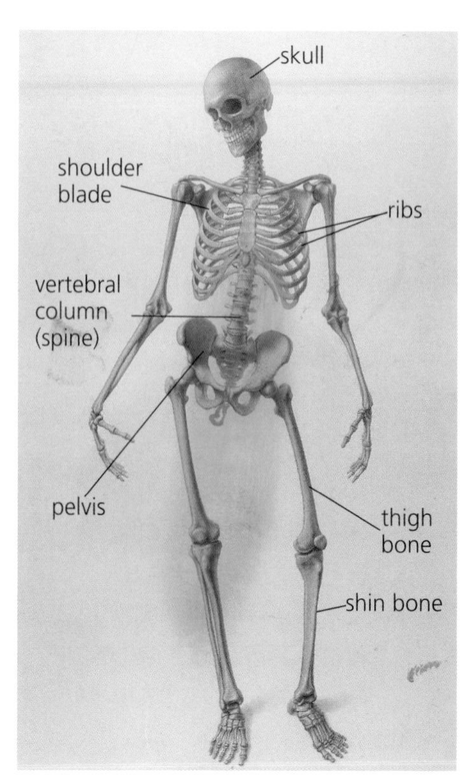

Movement and joints

Bend your lower leg backwards and forwards. If your leg was made of just one bone, it would be impossible to bend it. Now look at your arm. Move it around from the shoulder.

At certain points in your skeleton there are **joints**. Joints allow the bones to move. There is a joint in your knee and a joint at your shoulder. Different types of joint let bones move in different ways.

- Hinge joints let the bone move only forwards and backwards, but not side-to-side. Your knee and elbow are good examples. The bones are arranged a little like a hinge.
- Ball-and-socket joints let the bone move in all directions. Your shoulder and hip are good examples. On the end of one of the bones is a ball, which rotates inside a socket in the other bone.

You can see diagrams of these joints opposite. As you can see, all joints share many of the same features.

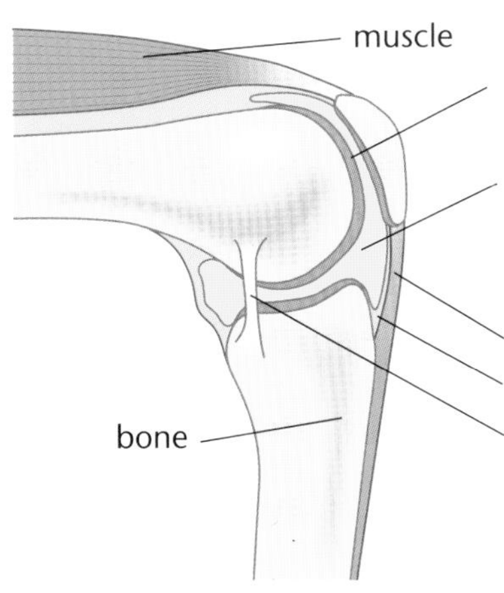

Hinge joint (knee).

Cartilage – smooth tissue which reduces friction, helping the ends of the bones slide over each other.
Synovial fluid – acts like a lubricant, reducing friction and helping the bones slide over each other. Also acts as a shock absorber.
Tendon – connects the muscle to the bone.
Synovial membrane – produces synovial fluid.
Ligament – holds the bones together. Made from a strong, stringy tissue that doesn't stretch very much. However, you can **sprain** (stretch) a ligament if you don't warm up properly before exercise.

pelvic bone

leg bone

Ball-and-socket joint (hip). The muscle and tendon are not shown.

b **Why are cartilage and synovial fluid important? What would happen if we did not have them in our joints?**

When joints go wrong

Anyone who does a physical job or plays a lot of sport puts their body under stress. Marc is a physiotherapist who treats injured people. Look at his diary showing what he did today.

Called to Snowhill football pitch today. The goalkeeper had collapsed in pain. Looked as if the cartilage in her knee joint had been wearing away for some time. She finally tore it today. I think she'll need an operation to remove any loose bits of cartilage.

Later on, went to that new housing development on the Snowhill estate. A builder had a slipped disc. His mate asked me about a sprain he'd got playing rugby on Sunday. I suggested he warm up properly before every game.

Did you know?

Your backbone is made of lots of tiny bones called vertebrae. There are joints between these bones, which contain discs of cartilage. If you hurt your backbone, the discs of cartilage can slip out of place. This is called a slipped disc.

c **What is a sprain?**

d **What causes a slipped disc?**

Questions

1 Make a table with three headings: 'Type of joint', 'Example', and 'The way it moves'. Complete the table for the two types of joint.

2 **a** Give two reasons why we have a skeleton.

b What makes bones strong?

3 In a car, there are lots of pieces of metal that move. To allow them to move, there are joints between the pieces of metal.

a The pieces of metal in a car are held together by screws. What holds the bones together in a joint?

b The pieces of metal in a car's engine need oil to help them slide across each other. What does the same job in a human joint?

4 Ron was trying to lift a heavy box when he got bad back pain. The doctor said he had slipped a disc. Explain what this means.

For your notes:

- The **skeleton** helps you stay upright, helps you move, and protects important organs.
- Bones can be moved at **joints** and are held together by **ligaments**.
- The hinge joint and the ball-and-socket joint allow movement in different directions.
- Joints contain **cartilage** and **synovial fluid** which help the bones slide across each other.
- You can injure joints by **spraining** a ligament, or when cartilage wears away or becomes torn. You can also break bones.

B5 Extra injury time

Learn about:
- Muscles
- Avoiding injury

Muscles and tendons

Alex never warms up before exercise. Unfortunately, he's torn his Achilles tendon and can't walk. Tendons are special fibres that join muscles to bones. They can't stretch, but they can tear.

Put your hand on the front of your upper arm, and bend your elbow. You should feel your biceps muscle **contract**, or get shorter and fatter. When the muscle contracts it pulls the tendon, which pulls the bones in your lower arm upwards. Now straighten your arm. You should feel your biceps muscle **relax**, or get longer and thinner. A muscle can't pull or push anything when it relaxes.

a **Try putting your hand on the back of your upper arm now. Bend and straighten your elbow. When does the muscle at the back of the upper arm contract and relax?**

A muscle is a bit like a piece of string. You can pull on bones using muscles, but you cannot push them. Muscles always work together in **antagonistic pairs**. For example, in the upper arm the biceps muscle contracts to pull the lower arm in one direction, and the triceps muscle contracts to pull it back again.

Look at the pictures on the right. If the tendon that links the biceps to the lower arm bones was torn, you wouldn't be able to bend your arm any more. It would be very painful. The same thing happened when Alex tore his Achilles tendon. He can't bend his ankle any more, or walk properly.

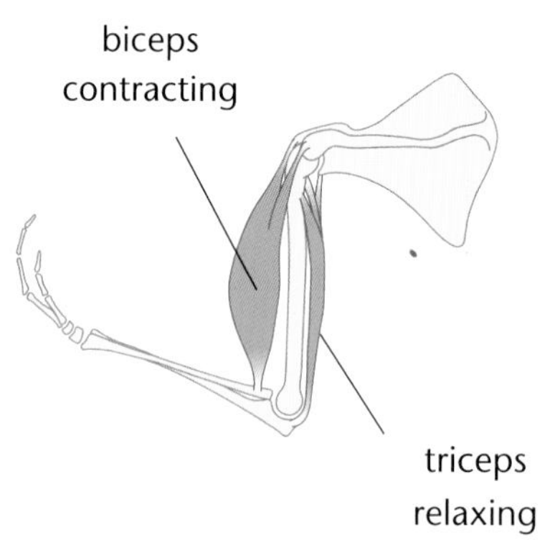

Raising the arm.

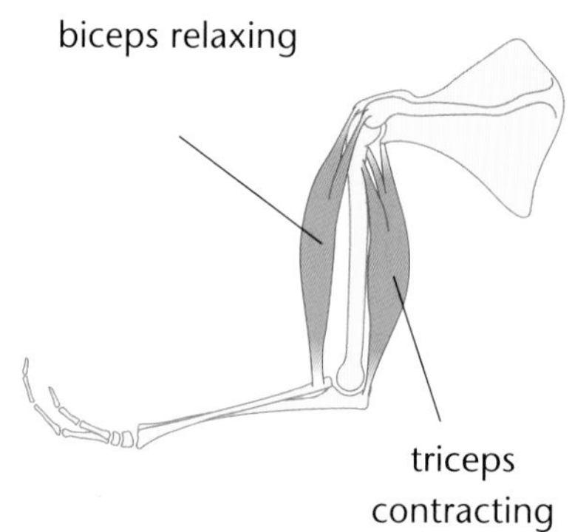

Lowering the arm.

b **Which muscle pulls on a bone to bend the elbow?**

c **Which muscle pulls on a bone to straighten the elbow?**

A muscle is an organ made of muscle tissue, along with capillaries to supply the muscle cells with blood. Muscle cells are very special because they can change shape. When a muscle contracts, all the cells contract. When a muscle relaxes, all the cells relax. If you look closely at a muscle you can see the muscle fibres as shown in this photo below.

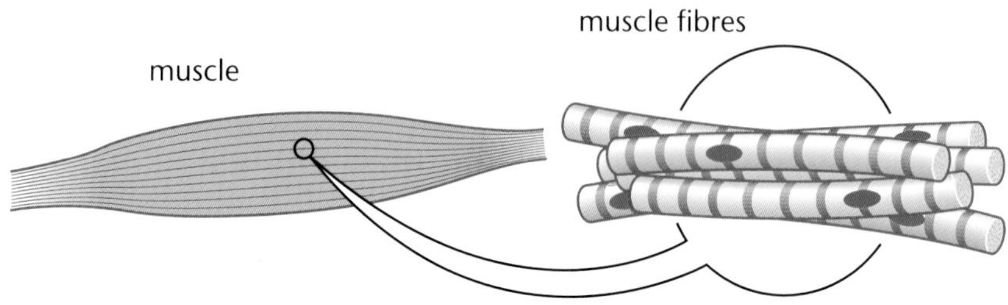

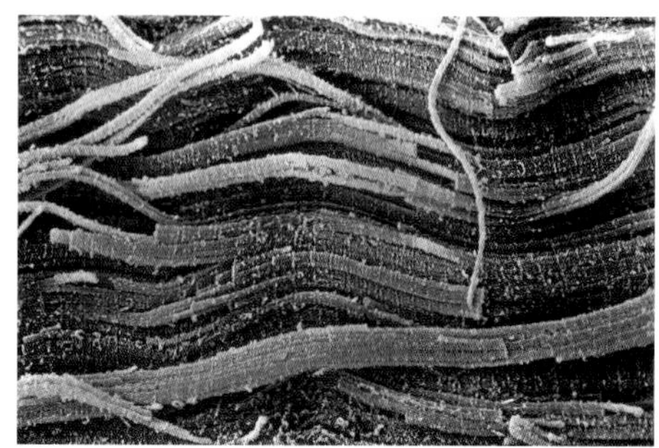

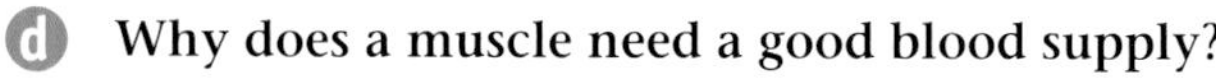

d **Why does a muscle need a good blood supply?**

Fitness programmes

All parts of our bodies need to work well together to keep us fit. Janet is a fitness instructor. She has made up a fitness programme for Ivor to help him lose weight and start getting fit.

Janet tells Ivor:

- To keep fit, you must keep your heart healthy to pump blood around your body.
- If your diet includes too much fatty food or alcohol, the blood vessels leading to your heart may get blocked. This stops the heart muscle getting enough glucose and oxygen for respiration, so it does not pump blood so well.
- Smoking increases your blood pressure, and makes your heart beat faster. This can wear out your heart.

Janet has also written a fitness programme for Mary, who wants to train for a marathon.

Janet tells Mary:

- Regular exercise increases the size of your heart and lungs, so that each heartbeat pumps more blood to your muscles.
- Regular exercise increases the number of capillaries carrying blood to your muscle cells.

Fitness programme for Mary

Continue to eat healthily

Swim 2 km two mornings a week

Run 5 km three evenings a week

Go for a very long run every weekend

e Suggest why the two fitness programmes are so different.

Janet always recommends doing a warm-up before beginning any physical exercise. It is easy to hurt muscles and joints when exercising. You may sprain a ligament or **strain** (pull) a muscle if you do not warm up properly.

Questions

1 Anne and Jessica went to an open evening at the local secondary school. In the biology department they measured the volume of their lungs, and measured their heartbeat. Anne had a bigger lung capacity than Jessica, and her heart beat more slowly. Who do you think exercises more regularly? Explain your answer.

2 Explain how muscles work together in antagonistic pairs to pull bones backwards and forwards.

3 Design a leaflet to explain to Year 7 pupils how to get fit. Explain why your suggestions will help.

For your notes:

- Muscles are joined to bones by tendons. Both can be injured if you do not warm up properly.
- Muscles contain muscle tissue. They can **contract** and **relax**. They work together in **antagonistic pairs** to move bones.
- To keep fit, you need to eat healthily, exercise sensibly and regularly, and avoid smoking and excessive drinking.

B6 Working together

Think about:

- How scientists work together

Faulty joints

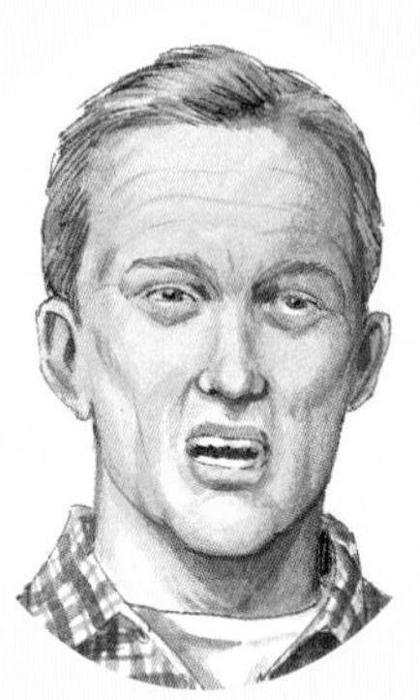

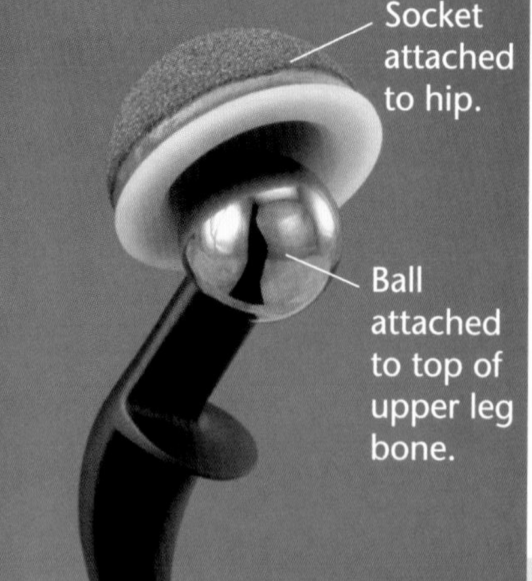

a **Look back at page 17. Which part of a joint could wear away and cause arthritis?**

You can see an artificial hip joint in the photo above. The socket is attached to the hip. The ball is attached to the leg, and rotates inside the socket. This is exactly how your hip joint works normally.

An artificial hip joint will get Reg active again and free of pain. But like his natural hip joint, the artificial one will wear out too. Artificial hip joints often need replacing.

Using evidence

Terry runs a lab researching artificial joints. He wants his team to design an artificial hip joint to last a lifetime. He knows that to make a really good joint, he'll need help from the technicians, engineers and materials scientists in his team. He'll also need expert advice from surgeons and physiotherapists who work with patients.

When scientists work together like this they often generate better ideas than when they work on their own. Terry called a meeting to look at how to improve the design of artificial hip joints.

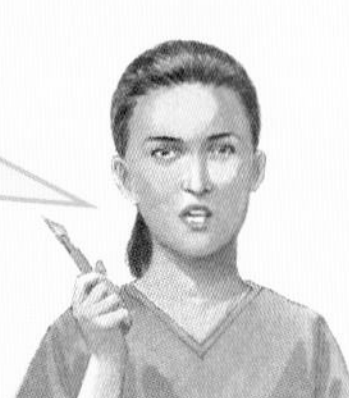

surgeon

physiotherapist

materials scientist

engineer

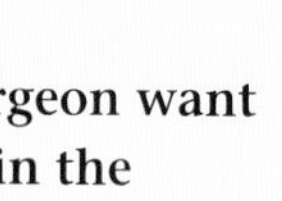

b **What does the surgeon want Terry to improve in the design of the joint?**

c **What does the physiotherapist want Terry to improve in the design of the joint?**

d **What does the materials scientist want Terry to change in the design of the joint?**

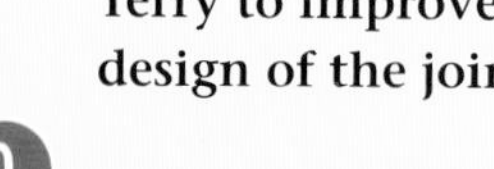

Terry decided to work on improving the way the socket is attached to the hip bone, and the material the ball and socket are made from.

Testing cement

Terry looked through medical journals where other scientists publish their findings. He found two ways of attaching the socket to the hip bone.

- You can use cement.
- You can make the hole in the hip bone exactly the same shape and size as the artificial socket, so the socket just slots in. When you do this, the hip bone actually grows into the joint, and holds it in place very firmly.

Terry asked the surgeon to find information about the failure of replacement joints shown in this table.

Age group	Type of fixing	Percentage of joints that failed within 10 years
20–39	cement no cement	25 8
40–59	cement no cement	16 17
60–79	cement no cement	8 25

e **Younger people's bones grow more quickly than older people's bones. Explain the difference in results between the 20–39 age group and the 60–79 age group.**

f **Which fixing method would you use for:**
(i) older people?
(ii) younger people?

Trialing materials

Terry did some research to find suitable materials for making his new hip joint. He decided to test the combinations in the table on the right.

The materials scientist used a hip-joint simulator to test the effect of movement on each one. The simulator moves the ball continuously backwards and forwards in the socket. For each combination of materials, he tested 50 artificial joints and found the average volume of material worn away in a year. You can see the results in this table.

Socket	Ball	Amount of joint worn away in a year, in mm^3
polyethylene	metal	2.8
metal	metal	0.8
ceramic	ceramic	0.004

g **Suggest two ways in which movement of the ball and socket in the simulator is different from movement in a person.**

h **Why did Terry test 50 hip joints for each combination of materials?**

i **Which materials would you choose for the socket and for the ball?**

Questions

1 Terry's team had to do trials with the simulator before the physiotherapist could get permission to trial the new joints in humans. Suggest why.

2 What do the engineer and material scientist need to do as a result of the trial findings?

3 Terry did trials with humans and finalised his new hip joint design. What materials and fixing method should he advise a surgeon to use for:

a an 18-year-old patient?

b an 80-year-old patient?

4 Look back through these two pages. Write a flow chart showing what you think each scientist did, and the order in which they did it.

C1 Hungry plants

Learn about:

- Photosynthesis

Where do plants get food?

Plants need food to respire and to grow, but how do they get this food?

I think that plants take in all their food through their roots.

I think that plants make their own food using energy from the Sun.

I think that plants make their own food from carbon dioxide and water.

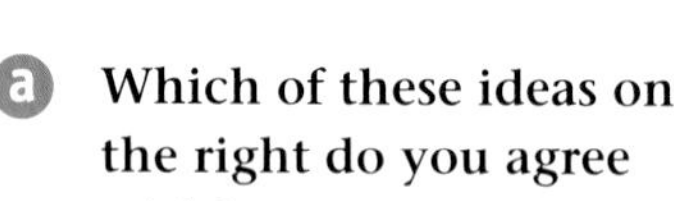

a **Which of these ideas on the right do you agree with?**

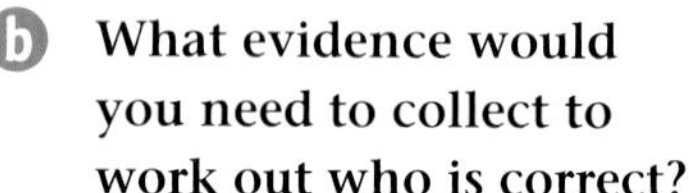

b **What evidence would you need to collect to work out who is correct?**

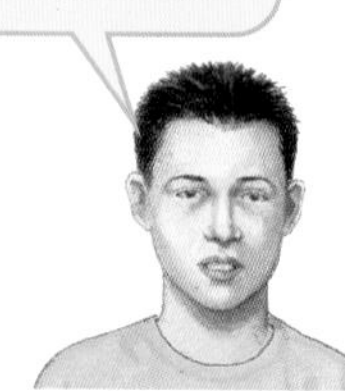

Matthew

Andrew

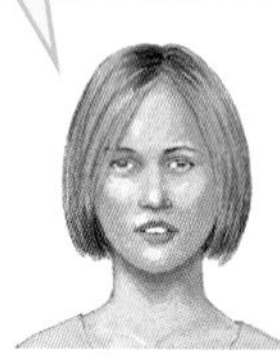

Jenny

In fact, both Andrew and Jenny are correct. Plants make their food in their leaves from carbon dioxide and water, using energy from the Sun. This process is called **photosynthesis**.

In the early 1600s, a Dutch scientist called Jan van Helmont carried out an experiment with a willow tree. Van Helmont filled a tub with 90.72 kg of soil, which had been dried in an oven. He planted a willow tree in the tub. The tree weighed 2.28 kg. To stop anything getting into the tub, he covered the surface of the soil. He kept it watered by adding only rainwater. Five years later he removed the tree and weighed it again; it weighed 77.51 kg. He then dried and weighed the soil; it weighed 90.67 kg.

c **Compare the increase in the mass of the tree to the decrease in the mass of the soil. What does this tell you?**

d **Where could the extra mass of the tree come from?**

How does photosynthesis work?

In photosynthesis, plants change carbon dioxide and water into oxygen and a sugar called glucose. They use the glucose for food. This change needs light energy from the Sun. Photosynthesis happens mainly in the leaves, though all the green parts of a plant can photosynthesise.

Did you know?

In the eighteenth century, the Dutch scientist Jan Ingenhousz was the first to discover that plants take in carbon dioxide and release oxygen when sunlight shines on them.

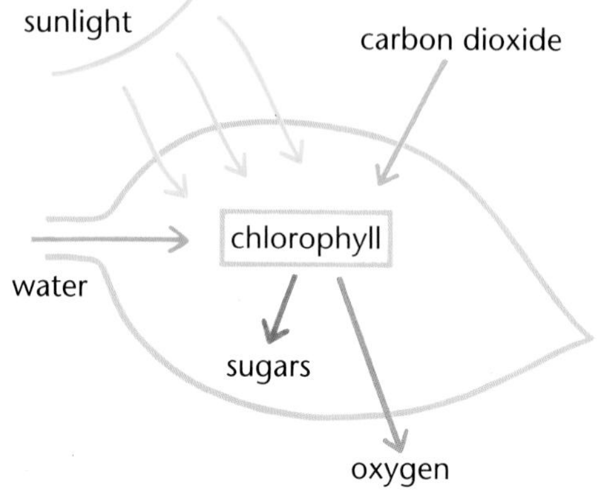

Jenny drew this diagram to show exactly what happens in photosynthesis. The light energy is transferred to the glucose as chemical energy. She wrote out the word equation and symbol equation.

$$\text{carbon dioxide} + \text{water} \xrightarrow{\text{light energy}} \text{glucose} + \text{oxygen}$$

$$6CO_2 + 6H_2O \xrightarrow{\text{light energy}} C_6H_{12}O_6 + 6O_2$$

How does light affect plants?

In Jessica's house there is a plant in every room. She thinks it makes the house look nice, but she can't work out why the plant in the living room grows more quickly than the plant in the hallway.

Using what you know about photosynthesis, answer this question for Jessica.

e **Which plant can photosynthesise most quickly? Explain why.**

Growing plants in a greenhouse

Anki lives in Holland. She grows flowers in greenhouses, and sells them all over the world. She wants to check at what times of day her plants are photosynthesising well. She can then supply extra carbon dioxide into the air in the greenhouse at these times, to help the plants photosynthesise even better. Many growers do this in their greenhouses so that the plants grow quickly.

Anki measured the carbon dioxide and oxygen levels in her greenhouses, and plotted these graphs on the right.

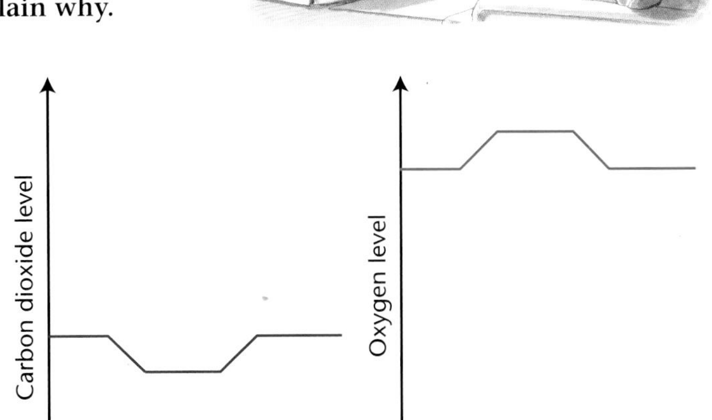

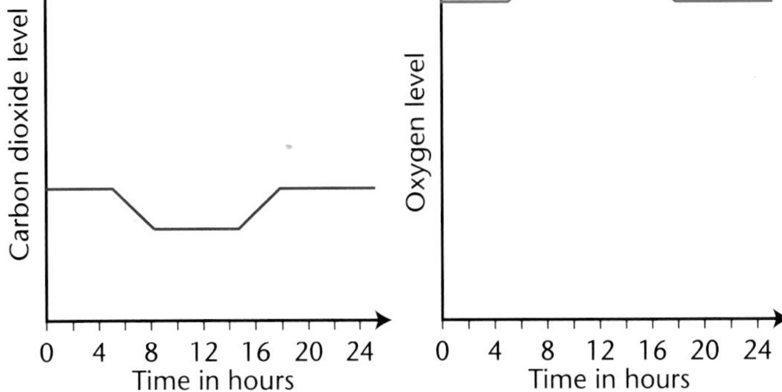

f **Between what times does photosynthesis stop? Explain how you worked this out.**

g **At what time of day is photosynthesis happening most quickly?**

h **When should Anki pump carbon dioxide into her greenhouses?**

Questions

1. Why do plants photosynthesise?
2. Look back at the equations for photosynthesis. Suggest two reasons why humans couldn't survive without plants.
3. Look at the experiment below. Tariq altered the distance between the lamp and the plant. He counted the number of bubbles of oxygen produced in one minute at each distance. He plotted his results on this graph.
   - **a** As the lamp got further away from the pondweed, what happened to the number of bubbles produced in one minute?

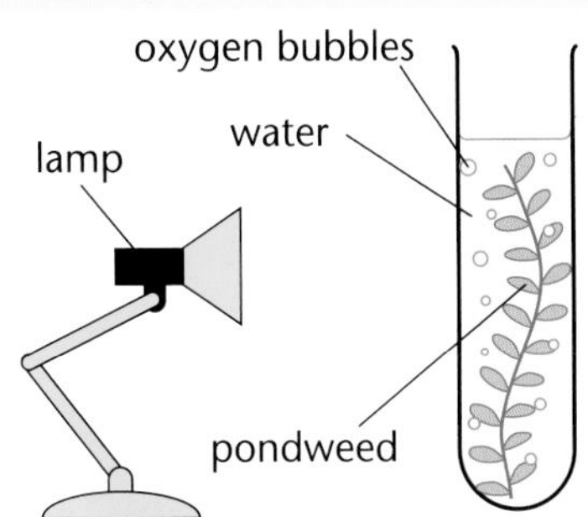

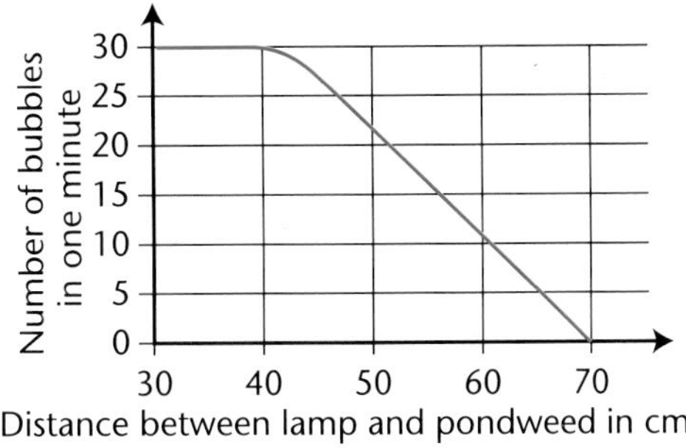

   - **b** As the lamp got further away from the pondweed, what happened to the speed of photosynthesis?
   - **c** Tariq studied his graph carefully. He realised that when the lamp is quite close to the plant, it doesn't matter how much closer you move the lamp, the speed of photosynthesis stays the same. Suggest a reason to explain this.

Did you know?

If you gave anyone flowers recently, it's very likely that they were grown in Holland. The Dutch are one of the biggest exporters of flowers in the world.

For your notes:

- Plants make food by a process called **photosynthesis**. This happens mainly in the leaves.
- In photosynthesis, plants take in carbon dioxide and water and use light energy to make oxygen and a sugar called glucose.

C2 A food factory

Learn about:
- Leaf structure

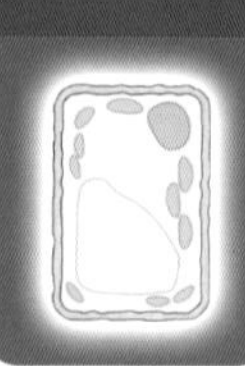

Catching the Sun

Look at the photo of the house. It is very 'eco-friendly'. Solar cells on the roof use light energy to produce electricity. They are electrical cells which make electricity when light hits them.

Solar cells are flat, and they have a large surface area. This makes them very good at taking in light. Because they are thin, light can reach every part in the cell.

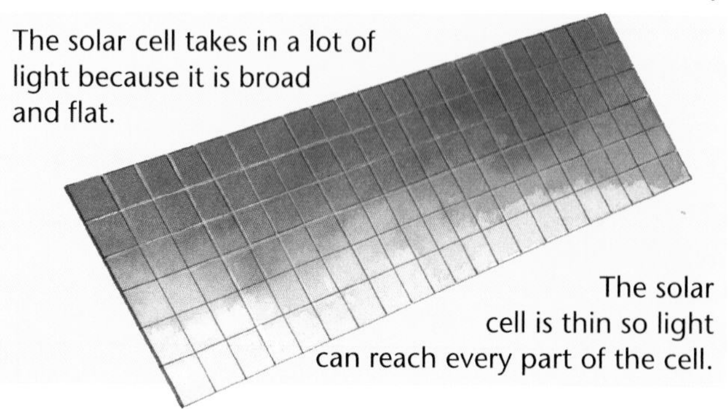

Looking at a leaf

A **leaf** is a plant organ that has a very similar function to a solar cell. Both of them need a lot of light energy to work. Plants need light energy for photosynthesis, which happens in the chloroplasts inside the cells of the leaf. Chloroplasts contain **chlorophyll**, a green pigment which takes in energy from the sunlight. All the green parts of a plant are able to photosynthesise to some extent because they contain chlorophyll.

Look at this photo of leaves.

a **In what two ways are solar cells and leaves similar?**

b **Why do you think that nettles growing in the shade often have broader leaves than nettles growing in the light?**

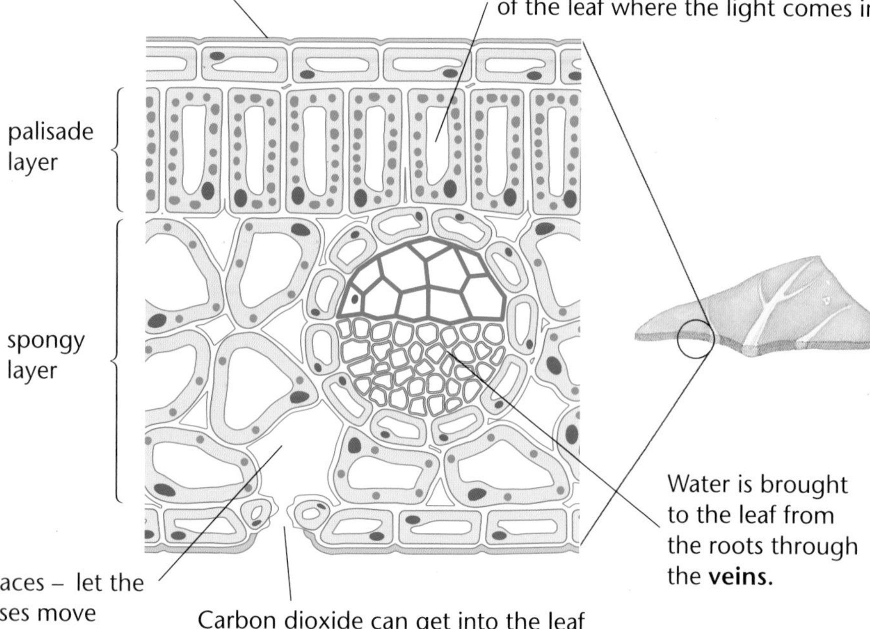

Looking close up

It's not just the shape of a leaf that is important. The cells of the leaf also help photosynthesis happen easily. We can slice a leaf open, and have a look at the cells inside it using a microscope. The cells are specialised to make photosynthesis as efficient as possible.

c **The top layer of cells is transparent. Why is this?**

d **What are the veins for?**

In any living thing, different cells have different functions. Cells are specialised to carry out their functions. This means they have special features to help them. Palisade cells are specialised for photosynthesis:

- They have lots of chloroplasts, which contain chlorophyll. Chlorophyll catches light energy, and so cells with lots of chlorophyll can photosynthesise well.
- They are long and thin. This means lots of palisade cells can pack tightly together, making sure that they catch as much light energy as possible.
- They are at the top of the leaf to make sure they get as much light as possible.

Gases in and out

Look at the leaf diagram again. Can you see the holes on the underside of the leaf? These holes are called stomata (one hole is a stoma). You can see them under the microscope in this photo.

The carbon dioxide for photosynthesis goes into the leaf through the stomata. The oxygen made by photosynthesis diffuses out of the palisade cells and into the spaces of the spongy layer before leaving the leaf through the stomata. The gases diffuse in and out of the leaf.

The cells on either side of a stoma are called guard cells. They are specialised to change shape to open or close each stoma. Look at the picture on the right below. The guard cells control the amount of carbon dioxide and oxygen that goes in and out of the leaf.

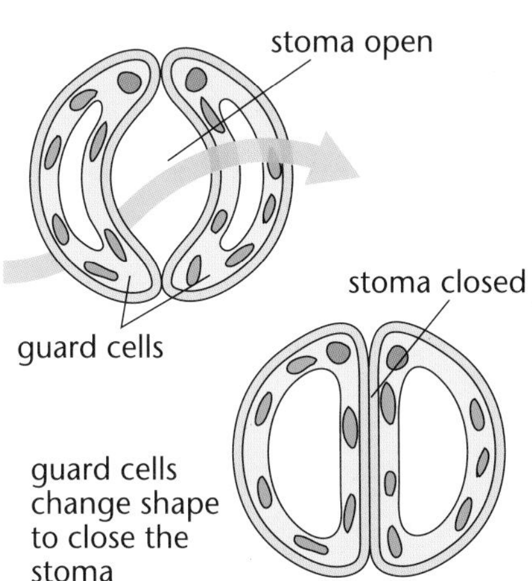

e Write down the parts of the leaf through which carbon dioxide travels on its way from the air outside the leaf to the chloroplasts in the palisade cells.

Questions

1 Write down as many similarities as you can think of between a leaf and a solar cell.

2 **a** How are the palisade cells specialised for photosynthesis?
b How does the shape of the leaf help with photosynthesis?
c What is the function of the air spaces in the spongy layer of the leaf?

3 **a** Explain the function of stomata in leaves.
b How are the guard cells specialised for their function?

4 Look at this picture of a variegated leaf. It has chlorophyll only in the cells towards the centre of the leaf.
a In which part of the leaf does photosynthesis happen?
b Suggest how the cells without chlorophyll stay alive if they cannot make their own food.

For your notes:

- **Leaves** are thin and have a large surface area, so every cell receives as much sunlight as possible.
- The structure of a leaf is adapted so it can photosynthesise well.
- The cells in a leaf are specialised to carry out particular functions to help with photosynthesis.

C3 Don't dry up!

Learn about:

- Water and nutrients
- Roots and root hairs
- Moving water around a plant

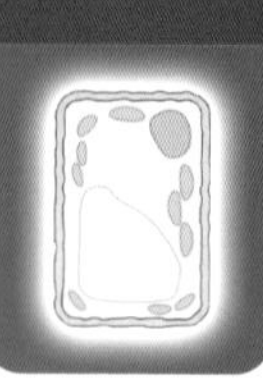

Why water?

Jamal's in trouble. His parents have got back from holiday, and their plants have drooped. Jamal was supposed to have watered them every day, but he didn't think it was important and he forgot.

a **What do plants use water for?**

What do roots do?

Roots are plant organs that grow deep into the soil. They have two main functions:

- They take in water and nutrients.
- They hold the plant firm in the soil.

Roots get thinner and thinner as they spread out. The very tips of roots have many tiny parts called **root hairs**.

Root hairs are long, thin parts of specialised cells called root hair cells. Root hairs stick out into the soil, giving a large surface area, which helps the roots take in water from the soil more quickly.

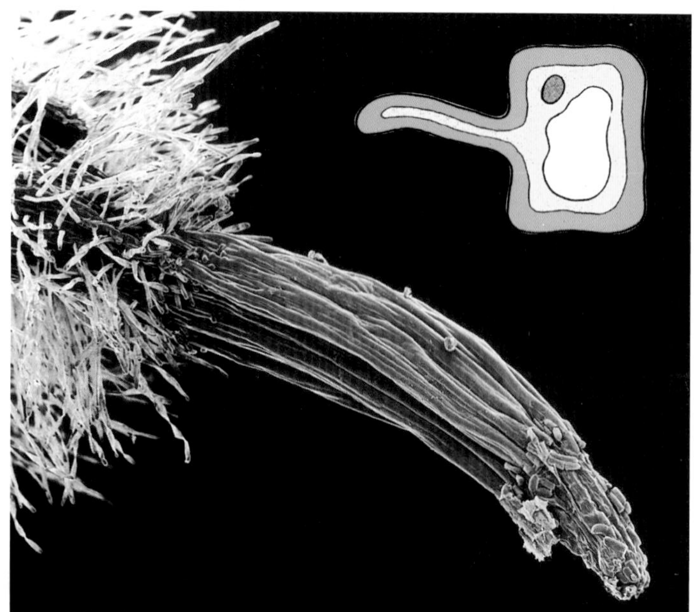

There are nutrients dissolved in the water in the soil, and the roots take in these nutrients along with the water. To increase the supply of nutrients, some gardeners also add **fertiliser** to the water. Fertiliser contains dissolved nitrates, phosphates and potassium, which all help to keep plants healthy.

b (i) **What is the function of root hair cells?**
(ii) **Explain how they are specialised to carry out their function.**

Reaching all parts

Once in the roots, water is carried up through the stem and into the leaves and flowers by tubes called veins. If you stand a plant in red dye, you can see the movement of water through the veins. If you cut a slice out of the stem, the red dye shows up in the veins. The celery stem in this photo shows this.

Did you know?

Some parts of a plant are almost all water. A ripe tomato is about 95% water.

All cells need water and glucose

Although plants need water, they don't need too much. If the roots have too much water around them, the root cells won't get any oxygen and will die. Just like other living cells, root cells need oxygen to respire.

Veins carry water and nutrients up the plant stem to the leaves, where the water is used for photosynthesis to make glucose.

If you don't water a plant, it wilts and goes floppy. Plants stay upright because every cell is full of water, keeping it rigid. You can see a cell from a healthy plant and a cell from a wilting plant in the picture on the right.

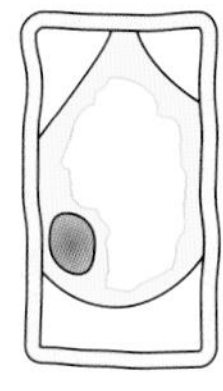

Cell from wilting plant.

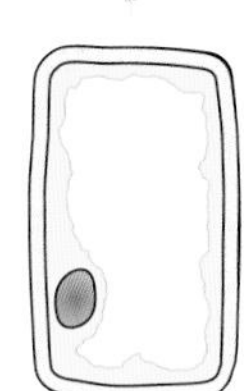

Cell from watered plant.

c What happens if you: (i) underwater (ii) overwater a plant?

Most glucose is made in the palisade cells, but all the plant's cells need glucose for respiration. To move around the plant, glucose is turned into sucrose. The sucrose dissolves in water and travels through the veins to other parts of the plant, where it can be used by the cells.

d Why does glucose need to be transported around the plant?

Questions

1 **a** Imagine you are a water droplet. Explain the route you take from the soil to a palisade cell in the leaf.

b Write down two reasons why a palisade cell needs water.

2 Maria was repotting a plant, and started examining its roots. Explain the answers to her questions:

a Why do roots spread out and get thinner and thinner?

b Why is it important for roots to have a large surface area?

c How do root hairs increase the surface area of roots?

For your notes:

- Plants need water and nutrients.
- **Roots** take in water and nutrients, and hold a plant firm in the ground.
- **Root hairs** have a large surface area to help roots take in a lot of water and nutrients.
- Water and nutrients are transported to all parts of a plant in the **veins**.

C4 Using plant biomass

Learn about:

- Biomass
- How we use plant biomass

Biopower

Look at the advertisement for the new electricity company BioPower.

a **What do normal power stations burn to generate electricity?**

b **What do BioPower power stations burn to generate electricity?**

Plant biomass is a renewable energy resource and burning it makes less pollution than burning coal or oil.

Do you realise that burning coal is polluting your environment?

Do you want to make electricity without burning coal?

Do you want to be environmentally friendly?

Then welcome to BioPower, the only electricity company to generate electricity by burning biomass.

BioPower

Biomass

Plants use glucose to respire and release energy to stay alive. Any spare glucose can be used to make other substances for the plant to grow. Plants convert glucose to cellulose, sucrose, fat and protein. Together these substances are used to make new cells and the plant grows.

All this new plant material is biomass. Cellulose is used to make cell walls. Fat goes into fruits like olives, which we use for making olive oil. There is protein in seeds and nuts.

Did you know?

Biomass is plant and animal material. It is a store of chemical energy. We measure biomass by finding the total mass of the animal or plant material, not including its water.

Most glucose is stored as starch immediately after it has been made. The picture on the right shows you how starch is made using glucose. We can test a potato for starch by adding a solution of iodine. The photo on the far right shows how the potato has gone black, showing starch is present.

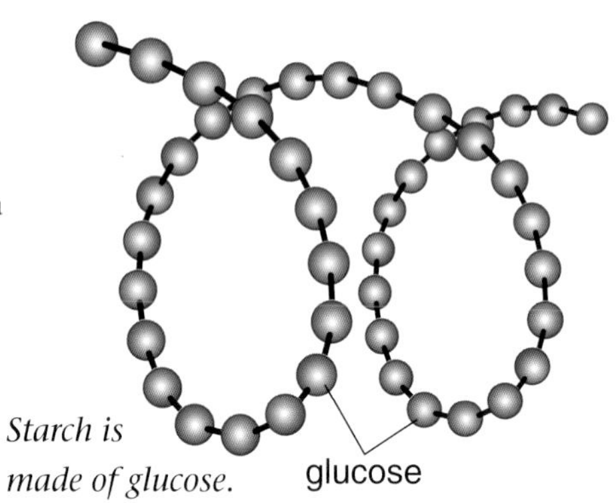

Starch is made of glucose.

Look at this picture of vegetables on a market stall. The plants make biomass in all the different plant parts, and some of them are good to eat. Many of these plant organs store starch.

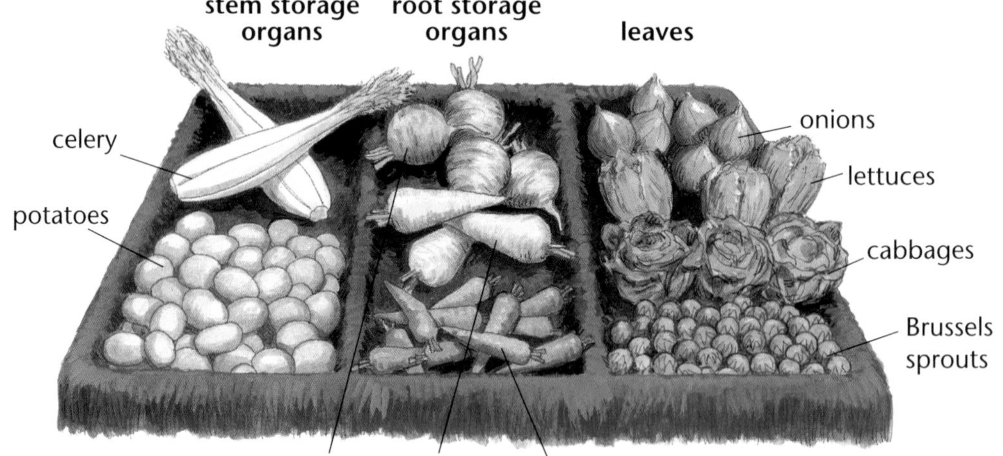

The carrot flower stores biomass in its seeds, ready to make new plants next year.

Plants can store starch in roots, stems and leaves to help them survive into the next year. If we did not pull carrots up and eat them, they would grow again in the spring. Plants also store biomass in fruit and seeds to help their seedlings grow and develop in the spring.

c Make a list of some of the nutrients humans need in a balanced diet that they can get from plants.

Using plant biomass

Look at these pictures showing how we use plant biomass. There are very many different uses for all parts of a plant – leaves, stems, flowers, seeds and roots.

Golden syrup is made from sugar cane. It contains sucrose.

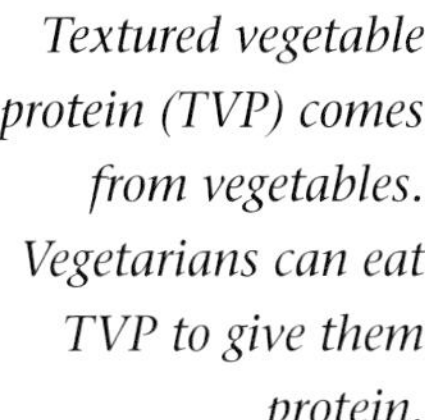

Textured vegetable protein (TVP) comes from vegetables. Vegetarians can eat TVP to give them protein.

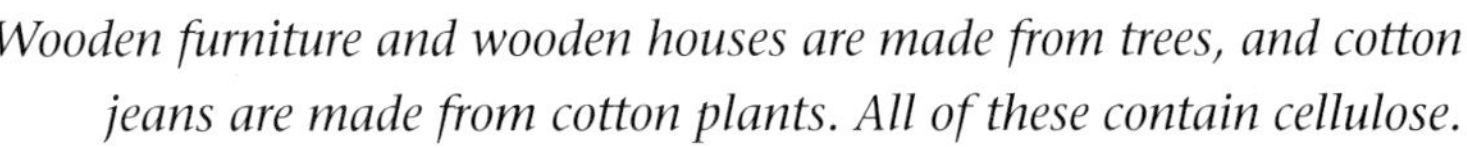

Wooden furniture and wooden houses are made from trees, and cotton jeans are made from cotton plants. All of these contain cellulose.

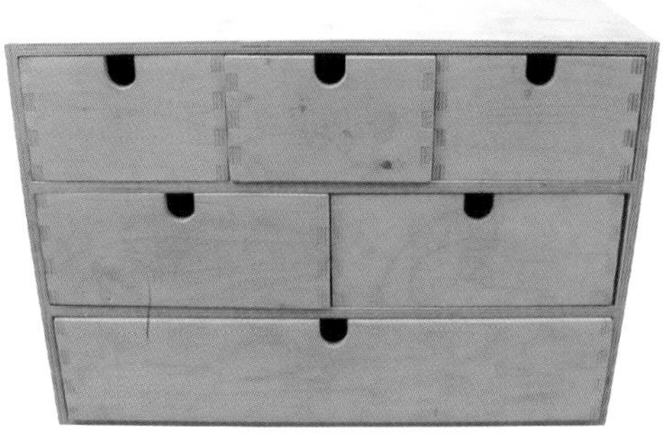

Corn oil is made from corn plants. It contains fat.

d Write down as many ways as you can in which humans use plant biomass.

Questions

1 Write a definition for the word biomass.

2 Why do plants store biomass?

3 Make a list of 20 objects in your house that are made from plant biomass.

4 **a** Name four substances that make up a plant's biomass

b Where did the energy originally come from to make these four substances?

5 Explain why and how starch is made in a plant.

For your notes:

- Biomass is the total mass of plant or animal material, not including water.
- Glucose can be converted into sucrose, starch, cellulose, proteins and fats.
- Humans use plant biomass in lots of different ways.

C5 Spot the difference

Learn about:

- Respiration and photosynthesis
- Green plants and the environment

Making food and getting energy

Fatima's younger brother Joe was asking her how plants eat. Having studied biology in Year 9, Fatima explained how plants photosynthesise to make their own food. She wrote down the word equation and symbol equation for photosynthesis, to make sure Joe would remember.

Photosynthesis

$$\text{carbon dioxide} + \text{water} \xrightarrow[\text{chlorophyll}]{\text{light energy}} \text{glucose} + \text{oxygen}$$

$$6CO_2 + 6H_2O \xrightarrow[\text{chlorophyll}]{\text{light energy}} C_6H_{12}O_6 + 6O_2$$

Fatima remembered that animals break down their food to get energy in respiration, and plants do the same thing. Plants respire to get the energy from the food they have made. She wrote down the word equation for respiration. The symbol equation for this reaction is shown underneath.

Respiration

$$\text{glucose} + \text{oxygen} \xrightarrow{\text{energy released}} \text{carbon dioxide} + \text{water}$$

$$C_6H_{12}O_6 + 6O_2 \xrightarrow{\text{energy released}} 6CO_2 + 6H_2O$$

a **Look at the two equations for photosynthesis and respiration. (i) What is similar about them? (ii) What is different?**

Day and night

Three pupils were discussing plant photosynthesis and respiration.

Plants photosynthesise during the day and respire only at night.

Plants photosynthesise during the day and respire all the time.

Plants photosynthesise at night and respire all the time.

b **Which idea do you think is right? Explain your answer.**

Look at the pictures to see what happens during the day and during the night.

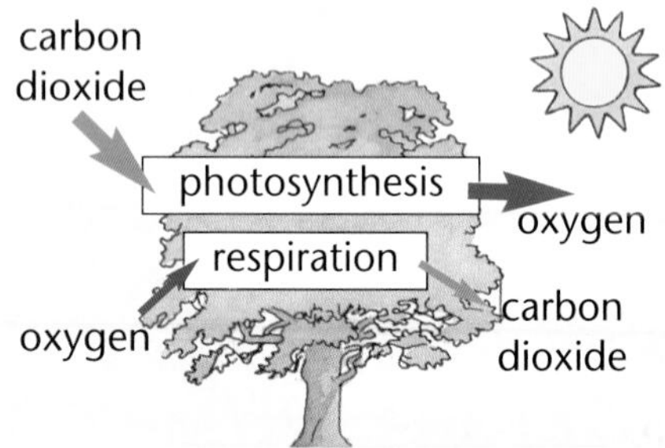

Photosynthesis is faster than respiration during the day.

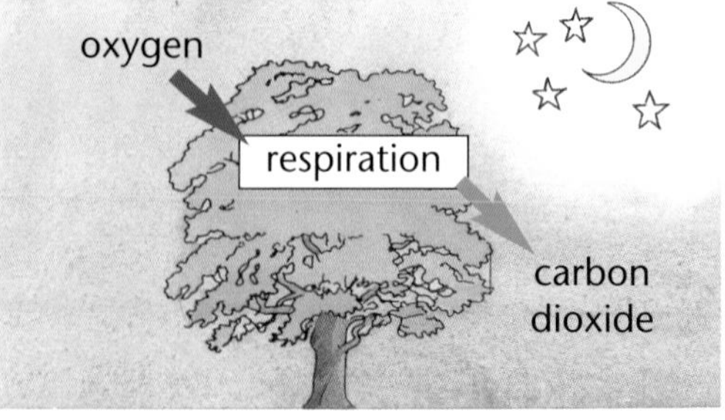

Only respiration takes place at night.

c **Which gas do plants make both during the day and at night?**

d **Which gas do plants make only during the day?**

In respiration, plants break down the glucose they have made in photosynthesis. Look at the equation for respiration again:

$$\text{glucose} + \text{oxygen} \xrightarrow{\text{energy released}} \text{carbon dioxide} + \text{water}$$

- Glucose comes from photosynthesis. Oxygen comes from photosynthesis during the day and from the air at night.
- Carbon dioxide is released through the stomata at night, and is used in photosynthesis during the day.

Why are rainforests important?

Look at this graph showing carbon dioxide levels in the atmosphere since 1700.

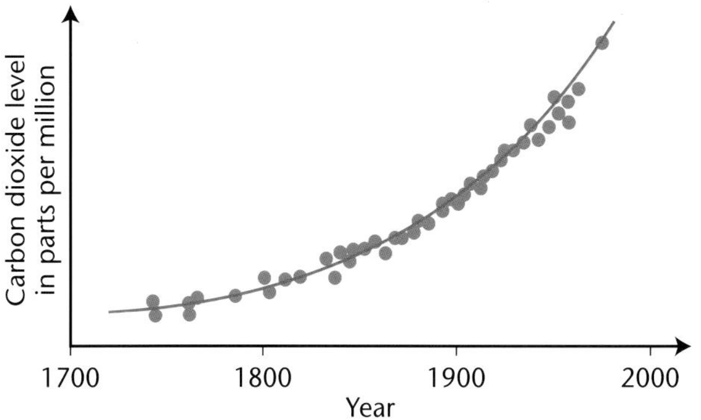

e **Describe how quickly carbon dioxide levels in the atmosphere went up in the 1700s compared with how quickly they went up in the 1900s.**

The rainforests are forests in tropical areas where it is warm and wet: conditions in which plants grow well. The rainforests cover huge areas of Africa, Asia and South America. Huge numbers of trees in the rainforests are being cut down, and scientists are very concerned about this.

Sometimes the forests are cleared so that the wood can be used for making paper, furniture or building materials. Sometimes the forests are just burned so the ground can be used for farming. People grow crops there which they can sell. The scientists are worried that all this tree felling will reduce the amount of photosynthesis, and carbon dioxide levels in the air will increase.

f **If there are fewer trees photosynthesising in the rainforests, explain what you would expect to happen to global oxygen levels.**

We all need oxygen for respiration, and we rely on plants to produce it by photosynthesis. More carbon dioxide in the air could increase global warming.

Questions

1 Students often find photosynthesis and respiration confusing. Mr Jones wrote down things that students said in his lessons when they were confused. Help Mr Jones by writing down why each student got it wrong.

a Respiration only happens in animals.

b Respiration happens in plants only during the night.

2 Copy this table and complete it using ticks and crosses to show which gases a plant gives out and takes in.

	Day	Night
gives out carbon dioxide	✓	✓
takes in carbon dioxide		
gives out oxygen		
takes in oxygen		

3 Plants are often taken out of hospital wards at night. Can you think of a reason for this?

4 Mary keeps a greenhouse of orchid plants. By thinking about respiration and photosynthesis, predict how carbon dioxide and oxygen levels in the air inside the greenhouse will vary over 24 hours.

5 The balance between photosynthesis and respiration is essential in maintaining levels of gases in the atmosphere. Explain what will happen to carbon dioxide and oxygen levels if:

a respiration happens more than photosynthesis

b photosynthesis happens more than respiration.

For your notes:

- Plants release energy from food by respiration, just like animals. Respiration takes place in every cell of a plant.
- Plants make their food, whereas animals must eat theirs.
- Cutting down rainforests may increase carbon dioxide levels in the air around the Earth.

D1 Storing food

Learn about:
- Food stores in plants
- Humans in a food web

Do you remember?

The equations for photosynthesis are:

carbon dioxide + water → glucose + oxygen

$$6CO_2 + 6H_2O \rightarrow C_6H_{12}O_6 + 6O_2$$

Best in show!

Marcus has entered a vegetable show. The rules are simple: the vegetable with the biggest mass wins!

Plants make biomass by photosynthesis. Glucose contains carbon, hydrogen and oxygen atoms from carbon dioxide and water molecules. You can see this in the symbol equation on the right. All the molecules in the plant except water make up biomass.

Why store food?

Plants use a lot of the glucose they make for respiration, but they store some as sugars and starch in different parts of the plant.

a **For each of the following plant parts, give one example of a plant that stores food there:
(i) stem (ii) roots (iii) fruit
(iv) seeds.**

It's tempting to think that plants store food for us to eat, but they really do it for the same reasons that people save money.

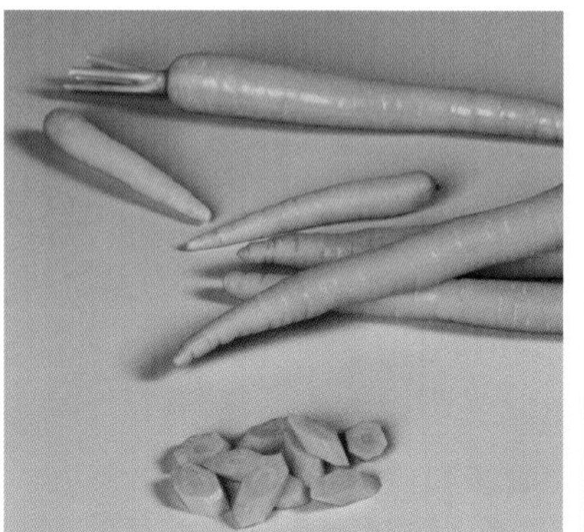

Why do people save money?

1. We save any spare money in the bank, ready for when we need it.
2. We save money to bring up children.
3. We put money in the bank to keep it safe from robbers.

Why do plants store food?

1. In the summer there's lots of light and plants photosynthesise more quickly than they respire. This means they make more glucose than they need, so they convert some to starch and store it. In the winter, when they need the glucose for respiration, they convert the starch back to glucose and respire it. Carrots, parsnips and beetroot are plants that do this.
2. In the autumn, plants store food in their seeds so that the tiny plant inside has food when it starts to grow in spring. Beans and peas are plants that do this. Plants also store sugars in fruit, to make it taste nice for animals. If an animal eats the fruit, it may carry the seeds inside to an area with good conditions for growth. Apples have seeds inside a sweet-tasting fruit.
3. Some plants like potatoes and rhubarb store food in their stems. Rhubarb also surrounds it's stems with poisonous leaves to stop animals eating them.

b **Explain why the winning vegetable in Marcus' vegetable show may not have the largest biomass. (*Hint:* look at what biomass means on page 28.)**

c **Explain why plants storing food in their seeds and fruit helps plant reproduction.**

Humans in a food web

Everyone in your class probably ate a different meal last night. Whatever meal you ate, the food in it came from animals or plants. Humans eat lots of different animals and plants, so they are in lots of different food chains.

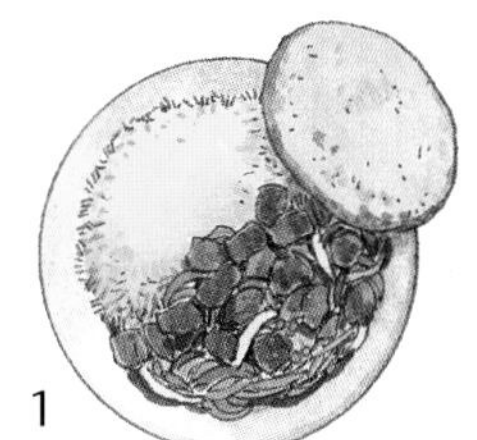
1

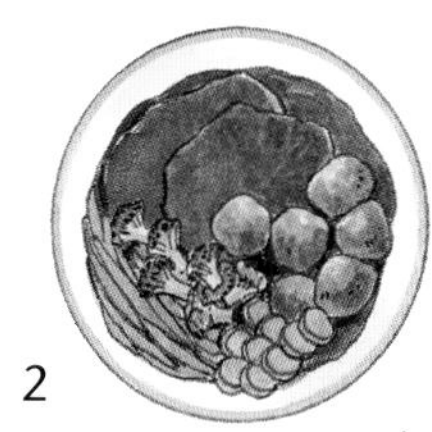
2

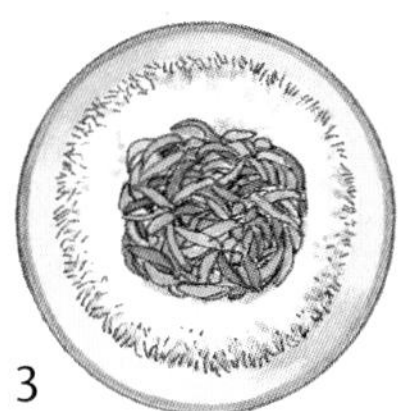
3

d **Write down a different food chain for each part of meal 2. For example, the food chain for the carrots would simply be:**

carrots → humans

If you put food chains together, you can make a food web. This shows how the energy is transferred through all the organisms in the food web.

e **Write out a food web that joins up all the different food chains you have identified.**

Look at this food web showing some of the animals and plants that humans eat.

Animals can't photosynthesise. They must eat other living things to get food for energy. They are consumers. When a consumer eats another organism, the energy stored in its biomass is passed on. The arrows show the energy transfer.

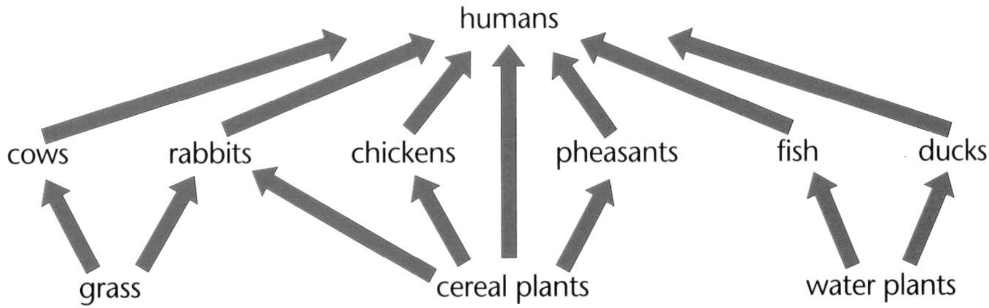

Plants photosynthesise using light energy from the Sun. The Sun is the energy source for the food web. Plants are producers.

Do you remember?

Energy enters food chains and food webs when the producers photosynthesise using light energy from the Sun.

f **How does energy enter the food web?**

There are three different types of consumer – omnivores, carnivores and herbivores.

Do you remember?

Humans are omnivores, we eat carnivores and herbivores.

Questions

1 a Write down three organs in which plants store food.

b List three reasons why plants store food.

2 Plants store food for similar reasons as we save money. Draw a table. In the left-hand column write out the reasons why we save money. In the right-hand column, match up the reasons why plants store food.

3 Write out as many food chains as possible from this food web.

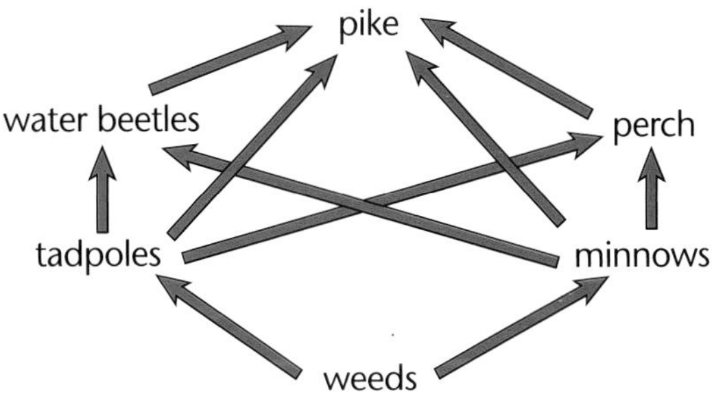

For your notes:

- Plants can store food in their roots, stem, fruit and seeds.
- Plants store food to help them survive the winter, and to help their seeds grow.
- Humans are in many food chains, and these join up to make a food web.

D2 Make them grow

Learn about:
- How plants grow best
- Nutrients and fertiliser

Clever greenhouses

When you buy someone flowers, they have probably been grown in a greenhouse like the one in this photo.

Greenhouses are specially designed to give perfect conditions for plants to photosynthesise and grow. Look at the greenhouse below.

a **What conditions do plants need to photosynthesise and grow well?**

Greenhouses are also used to grow crops. There are advantages and disadvantages to this.
Here are some advantages:
- Crops are healthy and look good.
- Crops grow for a longer season.
- Fewer weeds grow to compete with the crop.
- Fewer people are needed to look after the crop.

Here are some disadvantages:
- Some people think greenhouse crops don't taste as good as crops grown outdoors.
- It is expensive to buy the greenhouse and equipment.
- Because all the plants are close together, disease can spread easily.

Automatic shades control the amount of light. They let in enough light for photosynthesis, but not too much so that it gets too hot.

An automatic window and heater control the temperature. Photosynthesis slows down if it's too hot or too cold.

Carbon dioxide is added to the air in the greenhouse to help plants photosynthesise more.

A computer measures the water in the soil and switches water sprinklers on or off to stop the plants drying out or getting waterlogged.

Fertiliser is added to the water to give extra nutrients for growth.

When things go wrong

Sometimes the greenhouse equipment can go wrong. Look on the right at what happens when it does.

b **What happens to a plant that lacks:**
(i) carbon dioxide?
(ii) light?
(iii) water?
(iv) warmth?

The shades stayed across the window all day for a month, making it very dark in the greenhouse.

The carbon dioxide ran out and no one noticed.

Someone unplugged the computer that controls the water. The plants were without water for two weeks.

The heater stopped working at night, and the frost killed the plants.

What's in fertiliser?

Some plants need more nutrients than others. Some soils have very few nutrients. Farmers often add fertiliser to the soil to give extra nutrients to plants. Look at these fertiliser labels.

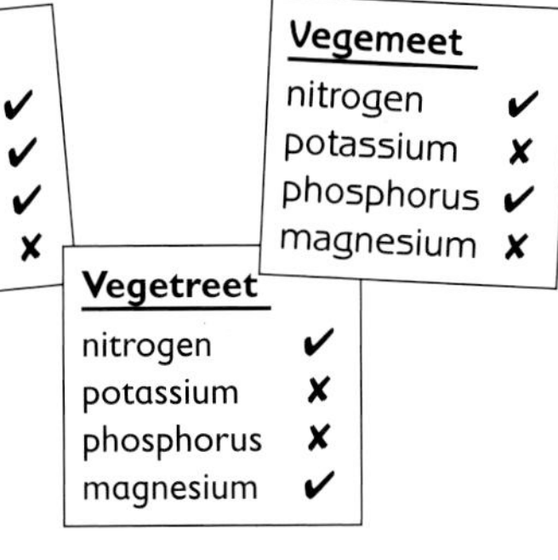

c **For what does a plant use: (i) phosphorus? (ii) potassium?**

If plants don't get enough of particular nutrients they can become unhealthly. You can see some of their **deficiency symptoms** in the pictures below.

Nutrient	nitrogen	phosphorus	potassium	magnesium
Why plants need it	Helps to make protein, which is important for growth.	Helps to make roots.	Helps to make flowers.	Helps to make chlorophyll.
What happens if plants are deficient	The plant grows slowly and has small pale leaves.	The roots and stem are short, and the leaves look purple.	Few flowers are produced. Leaf edges turn yellow and brown.	The plant looks yellow or brown and cannot photosynthesise well.

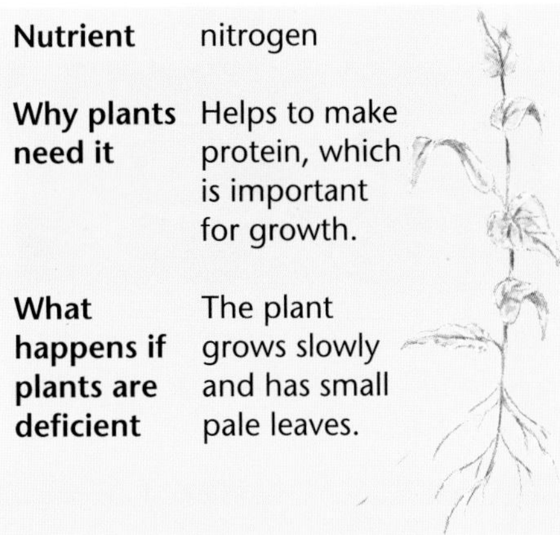

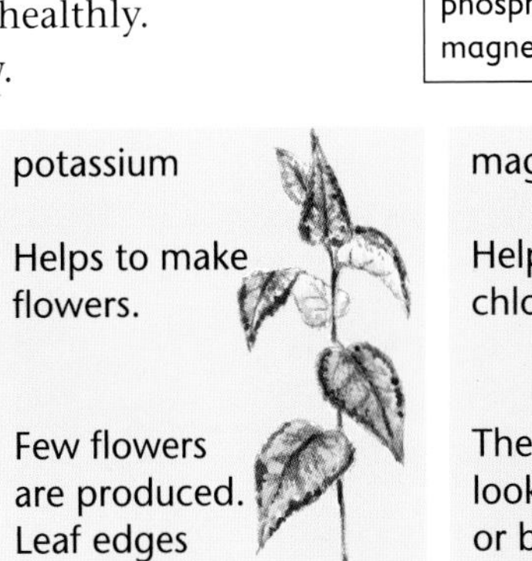

d **What happens if a plant does not get enough (i) nitrogen? (ii) magnesium?**

Fertiliser has to be dissolved in water so that the nutrients can pass through the root hairs. After they are absorbed, fertilisers stop plants getting deficiency symptoms, and help them to grow better. This means that farmers can make more money when they harvest and sell their crop.

Farmers like using fertiliser because it is easy to measure out the correct amount. The instructions give an **application rate**. This is the amount of fertiliser needed to fertilise $100 m^2$ of crop. Farmers can calculate how much it will cost to fertilise their fields using this information. The box on the right shows how they do this.

e **Mrs Moore has a field of grass that measures $800 m^2$. How much will it cost her to fertilise this field using the same fertiliser?**

The main disadvantage of using fertiliser is the cost. It costs money to buy it. It also costs money to spread it on the fields, because you need to pay someone to drive the tractor, and pay for the fuel. An alternative to fertiliser is manure which is free.

How much fertiliser to use

- A fertiliser costs £2 per kilogram.
- The application rate is 20 kg per $100 m^2$ of field.
- For a field of $200 m^2$ the cost is $2 \times 20 \times 2 = £80$.

Questions

1 Explain why plants need:

a water **b** magnesium **c** light.

2 Your father has just bought a greenhouse. He insists on closing the doors and windows very firmly, even in the summer. Explain to him why this is not good for his plants.

3 Bimla is worried about the plants in her garden. None of them are very healthy. Write a short guide for Bimla, so she can work out which nutrients her plants are missing.

For your notes:

- To grow well, plants need the right amount of light, water and carbon dioxide for photosynthesis.
- There are advantages and disadvantages to growing plants in greenhouses.
- Plants need nutrients to stay healthy and grow. **Fertilisers** provide plants with extra nutrients.

D3 Competing plants

Learn about:

- Competing with weeds
- Killing weeds

What do farmers think of weeds?

Weeds are plants that grow where you don't want them to grow. They need the same resources as crop plants, so they **compete** with them. If they were animals, they'd fight for the resources. But weeds compete with crops in less obvious ways.

a **List the adaptations of weeds which help them compete for light, water and nutrients.**

b **Look at the two pictures below. Which plants are the weeds in each picture?**

Weeds grow tall to catch the light before it reaches the plants below.

They also grow longer roots to take in water and nutrients from the soil before the other plants can get them.

Should we go organic?

Andrew used to work in the city, but moved to the countryside to become a farmer. He did not know anything about weeds, and did not do anything to get rid of them. Look at these bar charts. Andrew's crop yields were lower than those of his neighbours, so they gave him some advice. One neighbour, John, is an **organic** farmer – he avoids using artificial chemicals on his crops.

John says Andrew should pull up the weeds by hand.

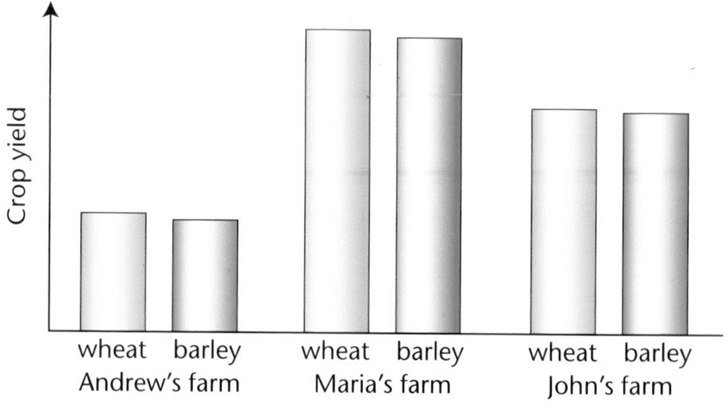

Maria says it's much easier and cheaper to use **weedkiller**. If Andrew pulled the weeds up by hand, he'd need about 20 people working for him.

c **Why do most farmers use weedkiller, rather than pulling the weeds up by hand?**

Most weedkillers are **selective**, which means they affect only the plant you want to kill. You can spray the weedkiller all over a field of crops, and it will only kill the weeds. Different weedkillers work in different ways. Some of them prevent the plant making protein so it cannot grow. Others make the weed grow very tall and thin, so eventually the stem breaks and the plant dies.

What about the environment?

Andrew thought about becoming an organic farmer, because he was worried that using weedkiller could upset the food webs on his farm. Farmers kill weeds to get a good crop. But killing weeds can have unwanted effects on food webs. Look at this diagram. It shows a food web in a sugar beet field. Fat hen is a weed that often grows in sugar beet fields. Skylarks, partridges and pheasants eat fat hen seeds.

Did you know?

Because organic farmers use no artificial chemicals, they have to employ more people to keep their farm running properly. This makes their produce more expensive.

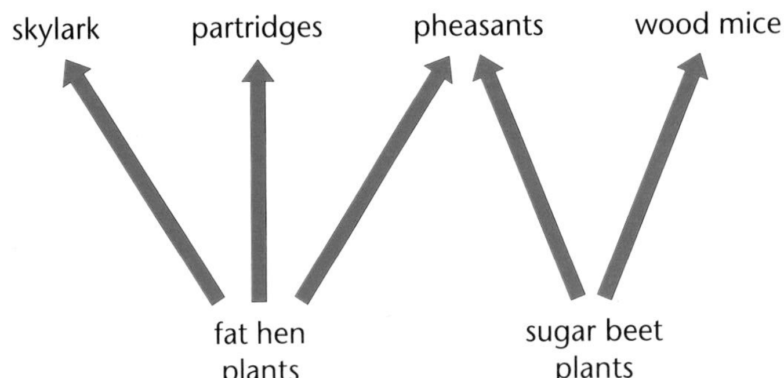

d **What would happen to the fat hen plant if weedkiller was sprayed on this field?**

e **What would happen to the number of skylarks and partridges?**

f **If the pheasants could not get food from fat hen plants, what would they eat for food?**

g **What would happen to the number of wood mice as a result?**

Farming and gardening can affect the number of different species that live in a particular area:

- Some farmers cut down hedgerows to give them more land to grow crops. This reduces nesting sites for birds, and the Government now pays farmers to plant new hedgerows.
- If farmers grow the same crop in every field, it will provide habitats for a very limited number of species. Any herbivores that don't eat this crop will have to go elsewhere for food, or they'll starve.
- You have probably seen garden makeover programmes on the television. Many people have low-maintenance gardens. A lawn or patio provides a home to only a few different species, whereas a garden with trees, hedges and shrubs can provide habitats for many species.

Questions

1 **a** List some ways in which weeds are specialised to compete with other plants.

b When dandelion plants grow in the shade of other plants, their leaves are broader than when they grow in the Sun. How does having broad leaves help them compete?

2 Organic farmers don't use artificial weedkillers. Use the information on this page to help you list:

a the advantages **b** the disadvantages of organic farming.

3 Make a leaflet advertising a weedkiller. Do some research using books or the Internet to decide which chemicals need to go into your weedkiller. Explain the effect these chemicals have on weeds.

For your notes:

- **Weeds** are plants that **compete** with other plants for light, water and nutrients.
- Farmers kill weeds using **weedkiller**.
- **Organic** farmers try to avoid using artificial chemicals.
- Removing one weed species from a food web can affect other members of the food web.

D4 What a pest!

Learn about:

- How pests affect plant growth
- How poisons build up in food chains

Who's a pest?

Field mice feed on farmers' crops. They compete with you directly for food. Animals like field mice, greenfly or slugs that feed on farmers' crops are called **pests**. They tend to invade a field in huge numbers because there is a good supply of food there for them.

birds of prey
humans
field mice
sparrows
wheat plants
barley plants

a Which animals in this food web are pests?

Did you know?

In Africa, plagues of millions of locusts can descend on a farm and eat every last bit of a farmer's crops. It's hard to compete against them!

These animals are all pests.

Getting rid of pests

Some farmers put out traps to kill field mice, because they eat the crop and cost money. Because most pests come in very big numbers, it's easier to use chemicals called **pesticides** to kill them. Many pests are insects, which can be killed by **insecticides**. A selective insecticide kills a particular type of insect. **Non-selective** insecticides kill any insect, whether it's a pest or not.

b What sort of insecticide do you think most farmers would use?

thrushes
bluetits
snails
insects
crop plants

What about the rest of the food web?

Insecticides kill insects which are a food supply for birds like bluetits and thrushes. The bluetits and thrushes are not pests, but they may die of starvation. Look at the food web on the left.

c If you used insecticide to kill the insects, what would happen to the population of bluetits and thrushes? (*Hint:* look at what else thrushes eat.)

The pyramids of numbers below show what happens to one food chain after spraying with insecticide.

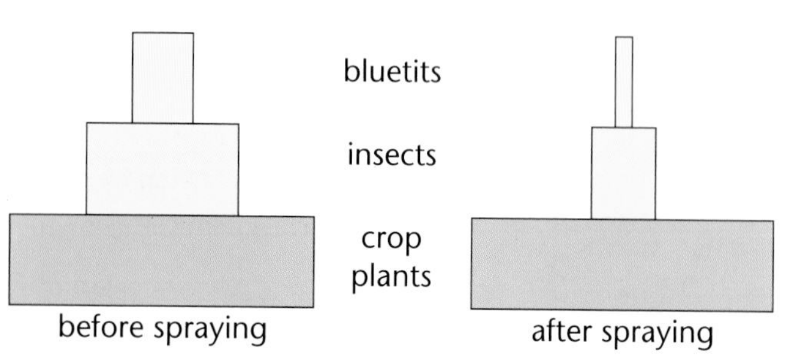

d Why is the bar for bluetits in the pyramid of numbers smaller after spraying?

Rose bushes are often plagued by greenfly. Ladybirds eat greenfly. If you spray a rose bush with a non-selective insecticide to kill greenfly, you will kill ladybirds as well. After you've sprayed, more greenfly may move into your garden from the garden next door. If there are no ladybirds ready to eat them, you may end up with more greenfly than you had before! Some gardeners buy ladybirds, or make sure they provide habitats to encourage them into the garden.

e Rather than kill the greenfly with chemicals, how else could you try to control their numbers?

Insecticides in the food chain

Instead of using a non-selective insecticide, you could spray the rose bush with a selective insecticide designed to kill only greenfly. But even that can cause problems ...

1 Most of the greenfly die. But some absorb only a little insecticide, and survive. The rose bush leaves absorb the insecticide as well.

2 As the surviving greenfly eat more rose bush leaves, they eat more and more insecticide. Insecticides contain poisonous substances called **toxins**. Some of these break down, but the ones that don't break down stay unchanged in the greenflies' bodies.

3 Bluetits eat greenfly. They often eat lots of greenfly, getting a tiny dose of toxin from each one.

4 Sparrowhawks eat bluetits. When a sparrowhawk eats lots of bluetits, it gets thousands of tiny doses of toxin.

Sparrowhawks are at the top end of the food chain. There are fewer bluetits than greenfly, and even fewer sparrowhawks than bluetits. There is the same amount of toxin, so it becomes more concentrated in the small number of sparrowhawks. The build-up of toxin poisons and kills the sparrowhawks.

Heron populations fall

Danger: DDT

After the Second World War, an insecticide called DDT was used to kill mosquitoes in southern Europe because mosquitoes spread the disease malaria. DDT killed all the mosquitoes and stopped the spread of malaria. About 10 years later, the heron populations started to decrease, and no one could work out why.

Scientists studied what was happening, and realised that DDT had built up in the herons. It did not kill them, but it was making their egg shells very thin. Every time a heron sat on her eggs, they smashed, so no young were hatched.

Organic alternatives

Instead of using insecticides, there are other ways to get rid of pests, especially in a greenhouse. Most pests have natural predators that eat them, and growers often use these to control the pests.

- Whitefly are pests that are eaten by a certain type of wasp.
- Greenfly are pests that are eaten by ladybirds, shown in this photo.

If you release wasps or ladybirds into your greenhouse, they eat the pests and help protect your crops from being eaten.

Questions

1 **a** What is a pest?

b Why do farmers kill pests?

c What unexpected problems can happen when farmers kill pests?

2 What is the difference between a selective insecticide and a non-selective insecticide?

3 **a** What happens to the concentration of a toxin in individual organisms as you go along the food chain?

b Explain why toxins are dangerous in food webs.

For your notes:

- **Pests** are animals such as insects that eat farmers' crops.
- Killing pests may cause unexpected and unwanted effects in other parts of the food web.
- **Toxins** build up in food chains and poison organisms at the top end of the food chain.

Think about:

- Sampling

D5 How many?

Counting wild animals

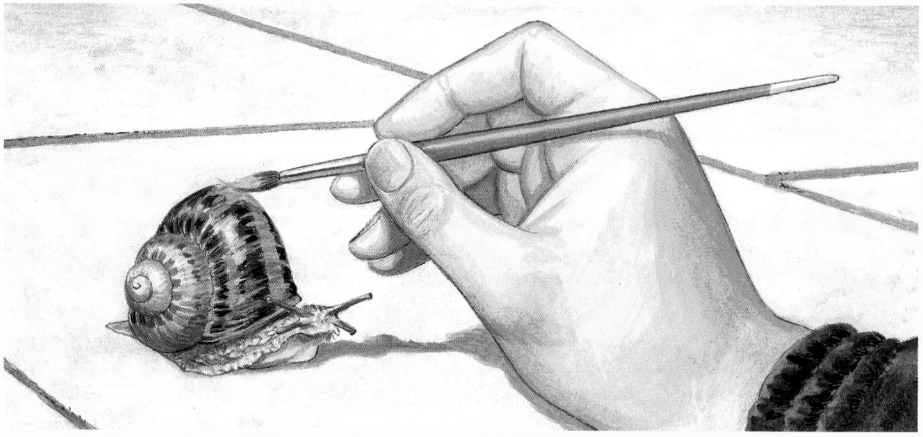

They caught 100 snails in traps. This was the first sample.

↓

They marked the 100 snails with paint.

↓

They released the snails.

↓

A week later they caught a second sample of 50 snails.

↓

They counted the snails in this sample with paint on their shells. There were 10.

James, Douglas and Christina have been watching a television programme about scientists studying a tropical rainforest. The scientists had found out about an endangered species of snail that lived on an island in the middle of a lake, in the middle of the forest. They wanted to know how many snails lived on the island, and so they sampled the snails. The flow chart on the right shows how they did this. From their samples they **estimated** the number of snails on the island (the population).

Ten of the snails in the second sample had paint on their shells. The scientists used ratios to estimate how many snails were in that part of the forest. Look at the way they worked it out.

total snails painted = 100
painted snails recaptured = 10
'painted snails recaptured':'total snails painted' = 10:100 = 1:10
The ratio means that the painted snails we recaptured make up one-tenth of all the painted snails. This means that for every one snail we catch, there are 10 times that number of painted snails still out there.
Just as the recaptured painted snails made up one-tenth of all the painted snails, we can say that the 50 snails we caught in the whole sample make up one-tenth of the whole population. To estimate the whole population, we need to multiply 50 by 10. That means there are 50 x 10 = 500 snails in the whole forest.

Let's have a look at how to write down what the scientists said using ratios:

- 'painted snails recaptured' : 'total snails painted' is the same as the ratio of 'size of second sample' : 'population'
- 'painted snails recaptured' : 'total snails painted' = 10 : 100 = 1 : 10
- 1 : 10 = 'size of second sample' : 'population' = 50 : population

You can write it like this:
10 : 100 = 1 : 10 = 50 : population

a **What was the ratio of 'painted snails recaptured' : 'total snails painted'?**

b **What was the ratio of 'size of second sample' : 'population'?**

c **On a different island, there were only 5 marked snails out of 50 in the second sample.**
  (i) What was the ratio of 'painted snails recaptured' : 'total snails painted'?
  (ii) What was the ratio of 'size of second sample' : 'population'?
  (iii) What is the population of snails on this different island?

Studying pollution

They caught 500 fish.

↓

They cut a notch in a fin on each fish.

↓

They released the fish.

↓

One month later they caught a second sample of 100 fish.

↓

They counted the number of fish with notches.

↓

They used ratios to estimate the population of fish.

Major oil companies have started drilling for oil in many rainforests around the world. Unfortunately, when they drill for oil, they sometimes release polluted water into the environment. An oil company built an oil-drilling platform at the lake where the snails lived. They started drilling in 1988 and kept going until 1996. The scientists wanted to know whether polluted water could affect one of the species of fish living in the lake.

The scientists sampled the fish on the lake every two years. They used different-shaped notches each time. The flow chart shows what they did. They also did the same at a lake that was not near an oil-drilling platform. Their results are summarised in the table.

d **Copy this table and fill in the missing populations.**

Site	1988	1990	1992	1994	1996
1 oil-drilling platform	5 5:500 = 1:100 = 100:? ? = ...	23 23:500 = 1:21.7 = 100:? ? = ...	36 36:500 = 1:13.9 = 100:? ? = ...	48 48:500 = 1:10.4 = 100:? ? = ...	50 50:500 = 1:10 = 100:? ? = ...
2 no oil-drilling platform	5 5:500 = 1:100 = 100:? ? = ...	5 5:500 = 1:100 = 100:? ? = ...	5 5:500 = 1:100 = 100:? ? = ...	5 5:500 = 1:100 = 100:?. ? = ...	5 5:500 = 1:100 = 100:? ? = ...

e **Make a bar chart for each site. Put the year along the *x*-axis and the estimated population up the *y*-axis as shown on the right.**

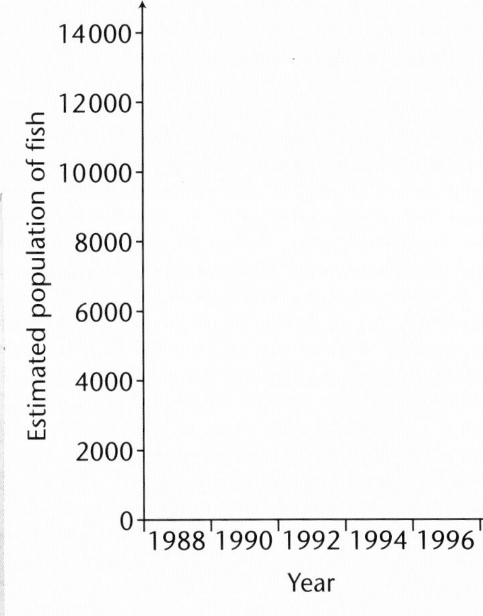

f **Does drilling for oil affect the population of this type of fish? Use the scientists' results in your answer.**

Sampling problems

James, Douglas and Christina were discussing what could have gone wrong with the snail-sampling experiment. Look at their notebook:

- The paint could make the snails more visible to predators.
- The snails could be poisoned by the paint.
- The shape of the trap may mean you only catch small snails.
- The painted snails may not mix with the rest of the snails in one week.
- There could be places on the island where few snails go, or where lots of snails go.

g **What effect would each of these problems have on the experiment?**

Questions

Think about the scientists' pollution research.

1 Why did the scientists study a lake that was not near an oil-drilling platform?

2 Suggest two things other than pollution that could change the population of fish.

3 The scientists thought that pollution from the oil was making the water acidic, which was killing the fish. What additional experiments should the scientists do to check their conclusions?

E1 What is a metal?

Learn about:

- The properties of metals

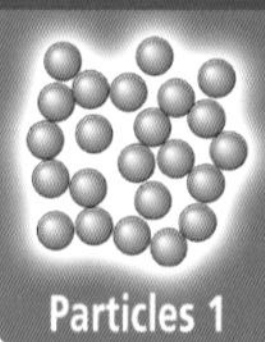

Are they useful?

Metals are very important in our lives. As you can see from this photo, lots of things around us are made of metals.

Metals have been important in people's lives throughout history. Stone Age people found pieces of gold in rivers before 5000 BC. The Egyptians used gold, silver, copper and bronze to make jewellery in 3000 BC.

Iron has been used in different ways over the years. The Hittites made iron from rocks in 1200 BC. The Chinese invented **cast iron** made by heating the rock with carbon at a very high temperature. In AD1665, Dud Dudley started to make cast iron to be used in industry.

Henry Bessemer made large amounts of cheap **steel** for industry in AD1856.

Did you know?

In AD 688 the Chinese built a pagoda of cast iron that was 90 metres tall!

Properties

Carly works as a courier for a mail order company. Her motorcycle of mainly made of metals. How a metal is used depends on how it behaves. We call this the metal's 'properties'. Look at the useful properties of the metals used in Carly's motorcycle.

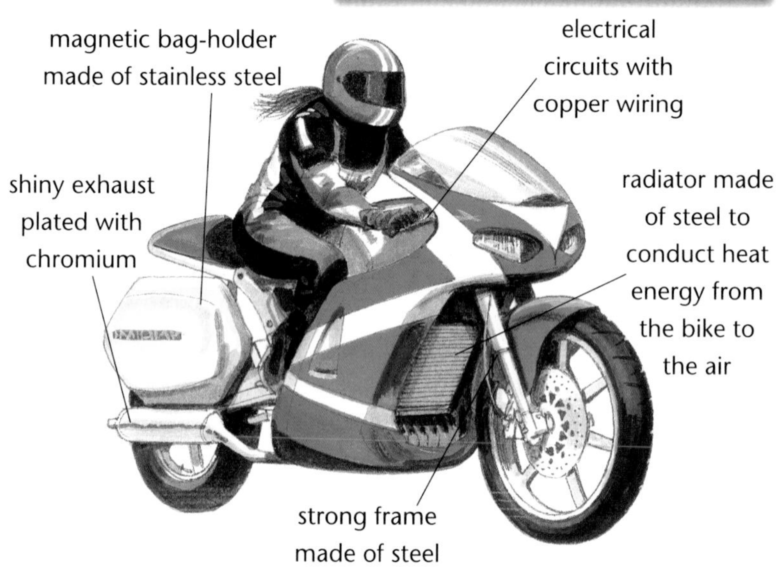

(a) Look at this picture. Which properties of metals do you think have been used in Carly's motorcycle?

Gold and silver are used for jewellery because they are shiny. Steel is used to make bridges and large buildings because it is strong. Iron is used for magnets, because it is magnetic. We discover the properties of a material by asking questions such as, 'Does electricity pass through it?' or 'What temperature does it melt at?'

Do metals conduct electricity?

You can use metals to complete a circuit because they let electricity pass through them: they conduct electricity.

Non-metals are poor conductors of electricity.

- Most wires are made of copper because it is one of the best conductors of electricity.
- Some wires, like those in the element of a toaster, are made of nichrome. This is a mixture of nickel and chromium. Nichrome can withstand high temperatures and has high resistance.
- The cables between electricity pylons are made of aluminium. It is not quite as good a conductor as copper but is cheaper and lighter.
- The prongs on most electrical plugs are made of brass. Brass is a mixture of copper and zinc. It is harder than copper and does not corrode.

Do metals conduct thermal energy?

We make saucepans of copper, cast iron, stainless steel and aluminium because metals let thermal (heat) energy pass easily: they conduct thermal energy. Non-metals are poor conductors of thermal energy.

What are they made of?

Metal things are either made of metallic elements or are mixtures. They can be mixtures of metallic elements or have non-metallic elements added. These mixtures are called **alloys**. Gold, silver, copper and iron are all elements. Brass is an alloy. Mixtures can be useful because they combine the useful properties of two materials. Brass is a mixture of copper and zinc. It is harder than copper and does not corrode like zinc, so it is useful for making door and window handles.

Do you remember?

Elements contain only one sort of atom. In the periodic table, the elements that have similar properties are grouped together.

Solid, liquid or gas?

Most metals are solids at room temperature (25 °C) – the temperature we are comfortable living at. Mercury is the only metal which isn't a solid at room temperature. Non-metals can be solids, liquids or gases.

b **Which metal would melt if heated from 25 °C to 100 °C?**

c **Which metal would not melt in the hottest part of a Bunsen burner flame (1500 °C)?**

d **Which three metals would boil in the hottest part of a Bunsen burner flame?**

e **Many thermometers contain mercury. Mercury thermometers can only be used between –38 °C and 350 °C. Use the information in the chart to explain why.**

Metal	Melting point in °C	Boiling point in °C
sodium	98	883
magnesium	650	1090
mercury	–39	356
calcium	850	1484
iron	1535	2750
copper	1080	2927
aluminium	660	2467
tin	232	2602
lead	327	1740
zinc	419	907
silver	961	2212

Physical and chemical properties

The properties of metals described on these pages are physical properties. In the rest of this unit, you will look at the chemical properties of metals and metal compounds, and how they react following some patterns. One example you have already met is that metals react with oxygen to make oxides.

Questions

1 Make a list of all the metals mentioned on these two pages. Give a use for each metal. Add any others you know. Which of these metals are alloys?
2 Draw a timeline to show how metals have been used throughout history.
3 Give one example in everyday life of a metal which is:
  a a poor electrical conductor   **b** a poor thermal conductor.
  Suggest what job each metal would be useful for and why.
4 Choose one of the metals mentioned on these pages and prepare a fact file on it, including its properties and main uses. Use poems or cartoons.
5 **a** Which metal is a liquid a room temperature?
  b How could you design a thermometer using this property?

For your notes:

- Metals are shiny and strong. A few metals, including iron, are magnetic.
- Metals are good conductors of electricity and thermal (heat) energy.
- Most metals are solids at room temperature.

E2 Salt on the roads

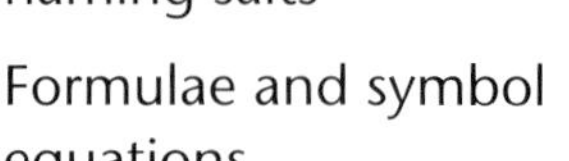

- Making and naming salts
- Formulae and symbol equations

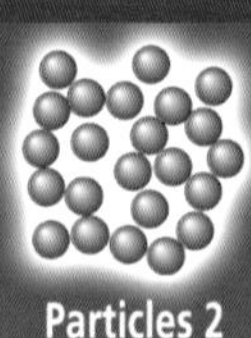

What is salt?

Every time icy roads are treated with salt, crystals of sodium chloride are sprinkled over them. In science, there are lots of different **salts** – sodium chloride is just one of them!

Salts are made by neutralising an acid with a base. Many bases are compounds of metals, such as metal hydroxides. Some metal hydroxides are soluble in water and are called alkalis.

Do you remember?

A base is the opposite of an acid; it cancels out acidity. An alkali is a soluble base; it makes an alkaline solution.

If you add vinegar (an acid) to sodium hydrogen carbonate (a base), you will see bubbles. The vinegar is used up. This is called neutralisation.

The reaction

When an acid is neutralised by an alkali, a salt and water are produced. In science, a salt is a substance we get from a neutralisation reaction. In the reaction between hydrochloric acid and sodium hydroxide, sodium chloride (common salt) and water are produced. The word equation for this reaction is:

sodium hydroxide + hydrochloric acid → sodium chloride + water

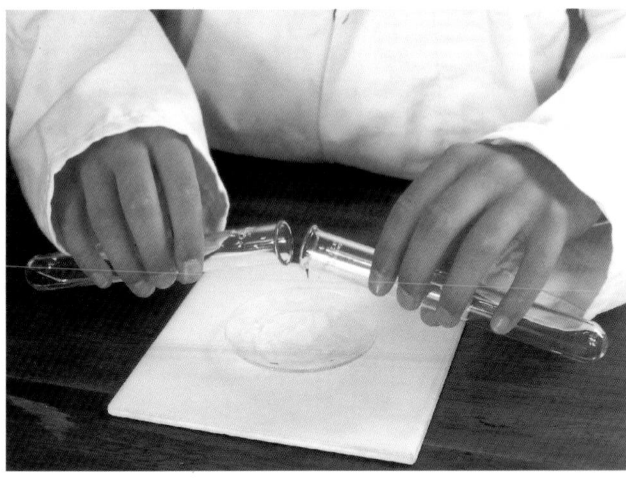

Making salts

If you take solutions of sodium hydroxide and hydrochloric acid that have the same concentration, and mix equal volumes of them, the sodium hydroxide neutralises the acid.

It is important to use the correct amounts of acid and alkali in a neutralisation reaction. You can use an indicator or a pH meter to test for neutralisation. If you test the solution at the exact point of neutralisation, the solution will be pH 7, which is neutral. (In fact, this is very difficult to do in practice.)

If you leave the solution to evaporate, you will see crystals of salt. This is the salt you put on the roads! There are many other useful salts made in the same way, with different names.

Compounds in short

We can represent a molecule of water using the formula H_2O. Symbols represent the atoms, and numbers tell us how many of each kind of atom there are in the compound. You cannot change the little number '2' in the formula for water. If you had a different number of hydrogen or oxygen atoms, the compound would not be water.

The formula for hydrochloric acid is HCl. It contains one atom of hydrogen for every atom of chlorine. We can use formulae to write symbol equations. For making common salt the symbol equation is:

$NaOH + HCl \rightarrow NaCl + H_2O$

Do you remember?

The number of each type of atom combining to form a particular compound is always the same. One atom of oxygen combines with two atoms of hydrogen to make one molecule of water.

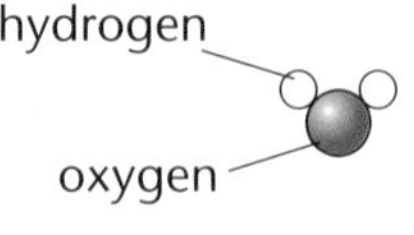

Datalogging

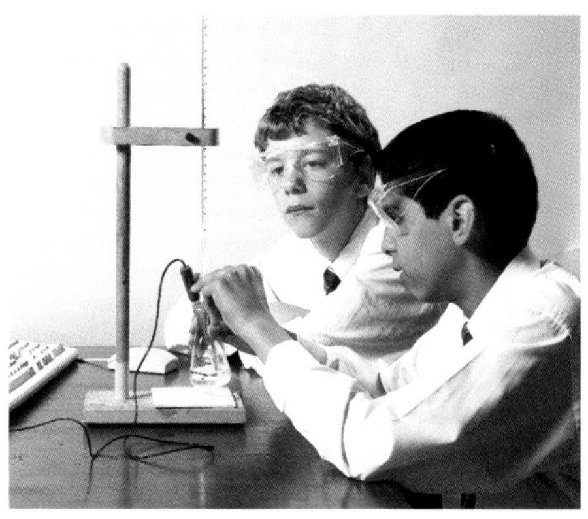

It is important to use the correct amounts and concentrations of acid and alkali in a neutralisation reaction. You can use an indicator or a pH meter to test for neutralisation. This photo shows Hassan and Martin using a pH sensor and datalogger to monitor the reaction.

At the start, they put 50 cm^3 of sodium hydroxide solution in the flask. They used a burette to add hydrochloric acid to the alkali, drop by drop, until the pH passed through 7. (It is very difficult to get exactly pH 7!) The volume of acid needed to neutralise 50 cm^3 of alkali was 5 cm^3. Then they continued to add the acid until the pH graph became level again. Here is the graph they displayed at the end of the reaction.

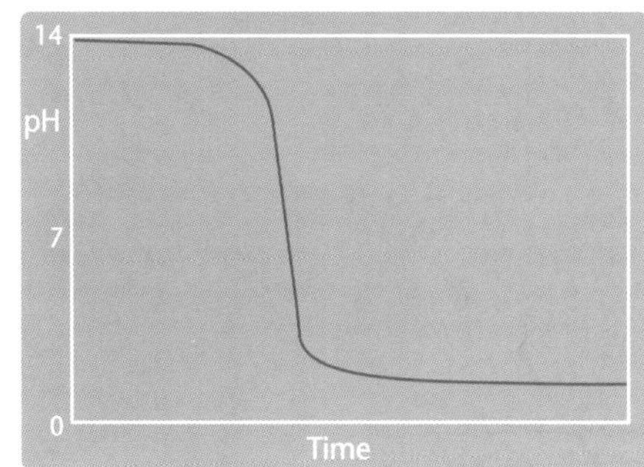

a **How did they know when neutralisation was complete?**

b **They concluded that the hydrochloric acid was 10 times more concentrated than the alkali. Explain how they worked this out from their results.**

Making other salts

If you use other acids and alkalis, you get different salts. You can write a general word equation for the neutralisation reaction:

acid + alkali → salt + water

A chemical reaction rearranges some of the atoms in the reactants to make products. Naming salts is similar. To name a salt, start by taking the name of the metal from the alkali. The second part comes from the acid – hydrochloric acid makes chloride salts. Nitric acid makes nitrate salts. Sulphuric acid makes sulphate salts.

The salt formed when you mix calcium hydroxide and hydrochloric acid is called calcium chloride.

Questions

1 Draw a flow chart to explain how to make common salt in the laboratory.

2 Design a poster to show what the products are in a neutralisation reaction between an acid and a base.

3 Write word equations for these reactions:
- **a** magnesium hydroxide + hydrochloric acid
- **b** zinc hydroxide + hydrochloric acid
- **c** calcium hydroxide + sulphuric acid
- **d** sodium hydroxide + nitric acid.

4 Which do you think is more accurate for monitoring a neutralisation reaction, an indicator or a pH sensor? Explain your answer.

5 Explain why the formula of water is always written H_2O.

6 **a** Soap contains a salt called sodium stearate. Stearates are made from stearic acid. The alkali used in soap manufacturing is sodium hydroxide. Write a word equation for this reaction.

b Shampoo contains a salt called sodium laureth sulphate. Which acid do you think is used in the manufacturing process?

For your notes:

- Alkalis neutralise acids to make a **salt** and water.
- The name of the salt comes from the names of the acid and the alkali used to make it.
- Hydrochloric acid makes chloride salts. Sulphuric acid makes sulphate salts. Nitric acid makes nitrate salts.

E3 Acids attack metals

Learn about:
- Metals and acids
- Balancing equations

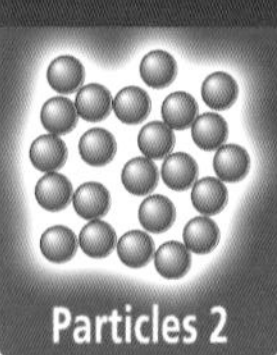

Disappearing metals?

The photo below shows part of an old car exhaust pipe. It has been corroded by the acidic gases from the exhaust.

Do you remember?

Metals are often corroded away by acids. They react to produce new substances.

Zinc and acids

If you add a few granules of zinc to some hydrochloric acid or sulphuric acid, it fizzes. The zinc is corroded and bubbles form. When all of the zinc has reacted, the bubbles stop.

a **How do you know that a chemical reaction has taken place?**

As well as hydrogen, the other products of the reactions are salts called zinc sulphate and zinc chloride.

The word equation for the reaction between zinc and sulphuric acid is:

zinc + sulphuric acid → zinc sulphate + hydrogen

The word equation for the reaction between zinc and hydrochloric acid is:

zinc + hydrochloric acid → zinc chloride + hydrogen

As with neutralisation, at the end of a 'metal + acid' reaction you may be left with a salt solution. Zinc sulphate and zinc chloride dissolve in the water from the acid. You can evaporate the water to leave the solid salt. Other salts may not dissolve in water. You can separate these by filtering.

You can write a general word equation for the reaction between metals and acids:

metal + acid → salt + hydrogen

All acids contain hydrogen. This is where the hydrogen comes from in the reactions with metals. Hydrochloric acid contains hydrogen and chlorine, and gets its name from both of these. Sulphuric acid also contains hydrogen. It's formula shows this: H_2SO_4.

Do you remember?

Some metals react with acids to make hydrogen gas. We can collect a test tube full of the gas and test it by putting a lighted splint near the top. Hydrogen gas is explosive, so you will hear a 'pop'!

b **Predict the name of the salt formed by reacting zinc and nitric acid.**

Symbol equations

Atoms are rearranged by chemical reactions, but the number of atoms in the reaction always stays the same. Look at this example on the right. Below the word equation are the formulae for the reactants and products and coloured dots for each type of atom. If you count the numbers of each type of atom on either side of the equation, the number is the same on both sides.

zinc +	sulphuric acid	→ zinc sulphate	+ hydrogen
Zn +	H_2SO_4	→ $ZnSO_4$	+ H_2
●	○○○●●●●	●○●●●●	○○

The symbol equation for this reaction is:

$$Zn + H_2SO_4 \rightarrow ZnSO_4 + H_2$$

It shows that the number of atoms on one side balances the number of atoms on the other side. We call this a **balanced equation**.

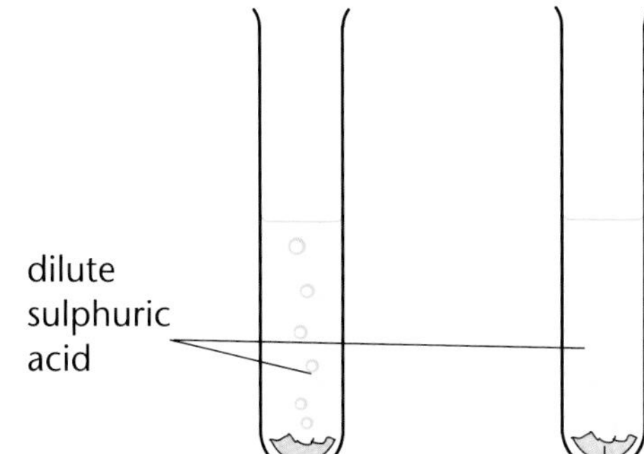

Different metals

Jade did an experiment to investigate what happens when iron and tin powders are added to dilute sulphuric acid.

C (i) **Which metal reacted most quickly with sulphuric acid?**
(ii) **What do the results tell you about iron?**
(iii) **Which metal did not react with sulphuric acid?**

Tin cans

Some cans for food and drinks are made of steel with a coating of tin on the inside. Others are made of aluminium. Tin reacts very slowly with dilute acids, so they are safe to use for food cans.

Aluminium reacts with oxygen in the air, forming a layer of aluminium oxide. This layer protects the aluminium from corrosion by acid. This protective layer means that aluminium can be used for storing acidic food.

Questions

1 What are the products of the reaction between a metal and an acid?

2 Write word equations for these reactions:
- **a** magnesium reacting with sulphuric acid
- **b** calcium reacting with hydrochloric acid
- **c** a reaction to make iron sulphate.

3 Predict what would happen if you put an iron nail into a beaker of hydrochloric acid. What two products would you end up with?

4 Hydrogen gas is produced when reacting zinc with sulphuric acid. Draw a diagram of the apparatus you would use to collect a test tube full of the gas.

5 You have three samples of colourless gases. You know that one is hydrogen, and the other two are oxygen and carbon dioxide. How would identify the hydrogen?

6 **a** Which metal and acid would you use to make common salt (sodium chloride, NaCl)?
b Write a word equation for the reaction.

For your notes:

- Many metals react with acids and are corroded away.
- When an acid reacts with a metal, a salt and hydrogen gas are produced.
- To test for hydrogen, put a lighted splint near the top of a test tube full of the gas. You will hear a 'pop'.
- We can represent chemical reactions using **balanced equations.**

E4 Acids attack carbonates

Learn about:
- Metal carbonates and acids

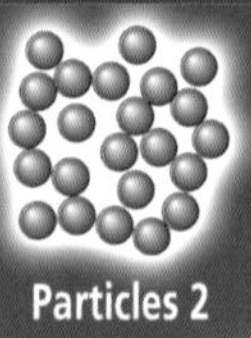

Limestone rocks

York Minster is built from limestone – mainly calcium carbonate. Over the years this has reacted with naturally acidic rain water, causing some of the surface of the building to be chemically weathered.

Do you remember?

Metals form compounds called carbonates. Acids react with metal carbonates. These are neutralisation reactions.

Indigestion

Your stomach makes hydrochloric acid, which helps to digest your food. Sometimes the stomach produces too much acid, so you get indigestion. Some indigestion tablets contain calcium carbonate. They fizz when you add them to acid. Calcium carbonate reacts with stomach acid to neutralise it. A salt and water are made, just as when any other base neutralises an acid. But there is also a third product – a gas.

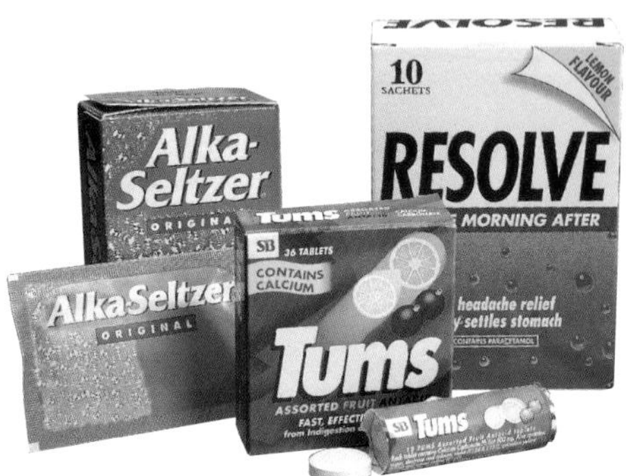

a **Which gas is produced when calcium carbonate reacts with acid?**

Carbonates and carbon dioxide

The word 'carbonate' tells us that this part of the compound has carbon and oxygen atoms in it. They stay together until the neutralisation reaction splits them up. They come together again to make carbon dioxide gas. The carbon and the oxygen atoms come from the carbonate.

Naming the products

To work out the name of the salt, take the name of the metal from the carbonate (calcium carbonate). Remember that hydrochloric acid makes chlorides. The name of the salt is calcium chloride.

The word equation for the reaction is:

calcium carbonate + hydrochloric acid → calcium chloride + water + carbon dioxide

You can write a general word equation for the reaction of metal carbonates with acids:

metal carbonate + acid → salt + water + carbon dioxide

Do you remember?

The test for carbon dioxide is to bubble the gas through limewater. The limewater turns milky.

b **Explain how atoms are rearranged in the reaction to produce carbon dioxide.**

Need a translation?

The Arabic text on the right describes how a new, fast-acting indigestion medicine works. It explains how the active ingredient, calcium carbonate, neutralises the excess hydrochloric acid in the stomach that causes uncomfortable heartburn.

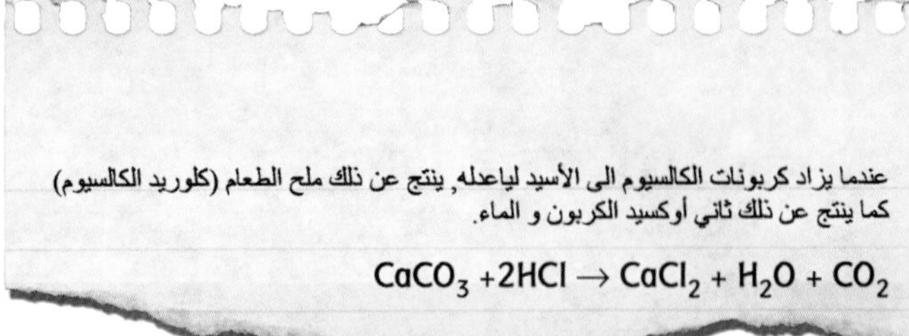

c **Explain why it is not necessary to speak or read Arabic to understand how this medicine works.**

Testing carbonates

This pupil set up an experiment to find out if other metal carbonates reacted with acids. He added some green copper carbonate to sulphuric acid and bubbled the gas through limewater.

He noticed:

- the reactants fizzed
- the colour changed to blue
- the limewater turned milky
- the boiling tube felt hotter.

Their first three observations showed that new materials were produced. The boiling tube getting hotter showed that energy was released. All these observations are evidence of a chemical reaction taking place.

d **(i) Which salt did he make?**
(ii) Write a word equation for the reaction.
(iii) Write a conclusion for their experiment.

Limescale

Look at the photo on the right of a kettle. The white substance that forms on the inside in some parts of the country is called limescale. It is calcium carbonate. Limescale also forms around toilet bowls. Some toilet cleaners have a weak acid in them, such as phosphoric acid, to remove the limescale.

The word equation for this reaction is:

calcium carbonate + phosphoric acid → calcium phosphate + water + carbon dioxide

Did you know?

Cola contains phosphoric acid. You could descale your kettle if you filled it up with cola!

Fire extinguishers

The reaction of an acid with a metal carbonate is used in red fire extinguishers. Inside the container is a concentrated sodium carbonate solution, a foaming agent and a glass bottle of sulphuric acid. Squeezing the trigger on the fire extinguisher breaks the glass bottle. The acid reacts with the carbonate to produce carbon dioxide gas, which forces the water out. The foaming solution of carbon dioxide in water puts out the fire.

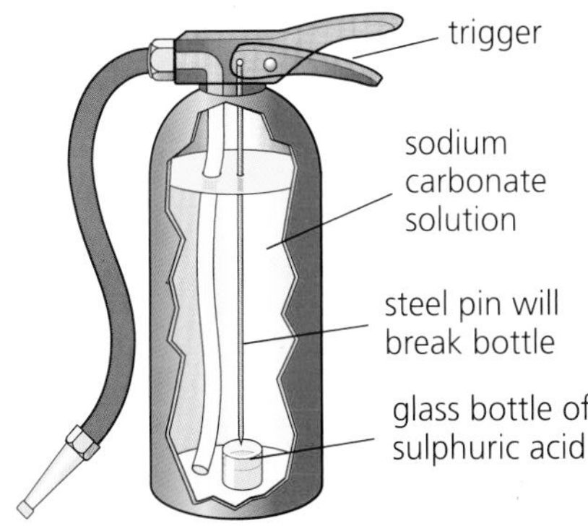

Questions

1 Explain why the fine sculptures on buildings such as York Minster are being worn away.

2 Why is it useful for chemical symbols and formulae to be the same all over the world?

3 **a** You want to find out whether copper carbonate reacts in the same way with hydrochloric acid as it does with sulphuric acid. Draw a labelled diagram of the apparatus you would use.
b What would you look for as evidence of a chemical reaction taking place?

4 You can descale a kettle with vinegar. It is a dilute solution of ethanoic acid, which forms salts called ethanoates. Write a word equation for this reaction.

5 Draw a diagram of a red fire extinguisher to show its internal design. Describe how it works.

6 Sodium nitrate is used in flares and as a fertiliser. It can be manufactured from sodium carbonate. Write a word equation for the reaction. Label the reactants and the products.

For your notes:

- Metal carbonates neutralise acids, producing a salt, water and carbon dioxide gas.
- The test for carbon dioxide is to bubble it through limewater. The limewater turns milky.
- New materials and energy changes are evidence of chemical reactions.

E5 Acids and metal oxides

Learn about:

- Metal oxides and acids
- Useful salts

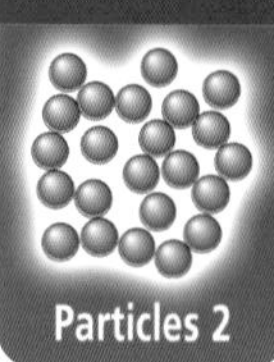

Particles 2

Stopping rust

Rust can be a problem on bikes and cars. The chemical name for rust is iron oxide. You can buy rust removers that contain acid. The acid reacts with iron oxide on the surface of iron or steel objects. This leaves a clean surface that you can paint.

Rust remover contains acid.

Do you remember?

You can separate a mixture of insoluble and soluble solids by dissolving it in water and then filtering it. You can separate a solute from a solution by evaporating the solvent.

Oxides reacting

Many metal oxides are bases. They neutralise acids. Black copper oxide reacts with dilute sulphuric acid if you heat it. A blue solution of copper sulphate is formed. If you leave the solution so the water evaporates, you will see blue crystals of copper sulphate. The crystals are a different colour to those you started with. This is evidence that a new substance is produced.

a **Look at the diagram. What process is used to:**
(i) separate any copper oxide that is left after the reaction?
(ii) get copper sulphate crystals from copper sulphate solution?

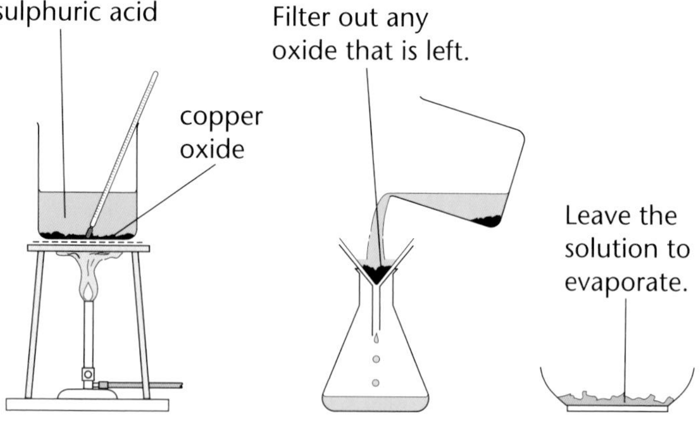

The word equation for the reaction is:

copper oxide + sulphuric acid → copper sulphate + water

The symbol equation is:

$CuO + H_2SO_4 \rightarrow CuSO_4 + H_2O$

We can write a general word equation for the reaction of metal oxides with acids:

metal oxide + acid → salt + water

Copper sulphate is a useful salt. It is used to spray grapevines to prevent disease.

Did you know?

Old copper coins look black because they are covered with black copper oxide. If you put them in vinegar (an acid), the copper oxide reacts with the acid to give you shiny coins in mint condition!

Copper sulphate crystals.

Time to investigate

Laxatives are medicines that relieve constipation. Some laxative medicines are made from magnesium sulphate. Lianne and Adam worked out that you could make magnesium sulphate by reacting magnesium oxide with sulphuric acid instead of copper oxide. They added some magnesium oxide to sulphuric acid. This time there was no colour change. They filtered the solution and evaporated the solution. White crystals of magnesium sulphate were left.

The word equation for the reaction is:

magnesium oxide + suphuric acid → magnesium sulphate + water

The coloured dots under this symbol equation show how many of each type of atom are involved in the reaction:

MgO	+	H_2SO_4	→	$MgSO_4$	+	H_2O
○●		○○○●●●●		○ ○●●●●		○○●

A magnesium sulphate crystal.

b **Explain how atoms are rearranged in the reaction to produce water.**

More useful salts

Many salts are useful compounds.

- Copper chloride is used to treat fish fungus.
- Iron sulphate is used in iron tablets. These are given to people suffering from anaemia.
- Silver nitrate is used in photography.
- Potassium nitrate is used in gunpowder.

c **Predict the name of the acid you would need to make each of these salts.**

Did you know?

Soap contains a salt called sodium stearate. It is made from sodium hydroxide and an animal fat which has stearic acid in it .

Questions

1 What evidence is there that a chemical reaction takes place when you add copper oxide to sulphuric acid?

2 **a** Look back at the formula of copper sulphate. How many different sorts of atoms does it contain?

b Which atoms appear to be going around as a group?

3 **a** Calcium sulphate is used to make plaster of Paris – the plaster you have if you break a bone. Name the metal oxide and the acid you would need to make calcium sulphate.

b Write a word equation for the reaction.

4 **a** Describe two uses of salts in medicines.

b Design a poster to explain how to work out the name of a salt.

5 Hydrochloric acid is used to prepare steel surfaces for galvanising. It removes the rust (iron oxide). Write a word equation for this reaction.

For your notes:

- Many metal oxides are bases. They react with acids, producing a salt and water.

F1 Losing that shine

Learn about:

- Metals reacting with oxygen

All that glitters ...

Gold is a beautiful shiny metal. It is perfect for jewellery but is very expensive. You can buy cheaper jewellery made from other metals. These look fine at first but soon lose their shine. Most metals lose their shine when they react with the oxygen in air to form a metal oxide.

Reactivity

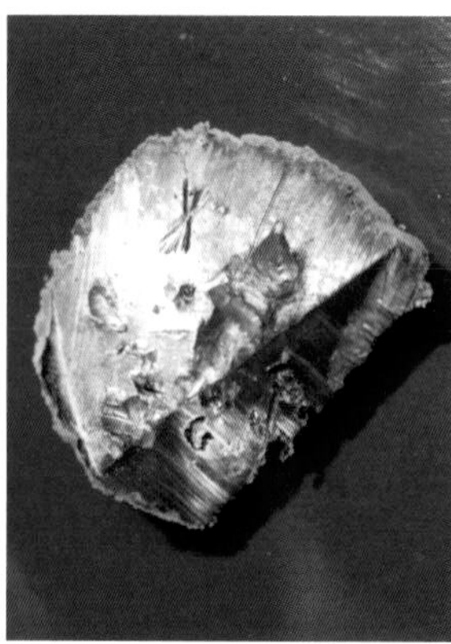

This potassium will lose its shine in seconds.

Gold is a very **unreactive** metal. Gold jewellery can last thousands of years without changing, because it does not react with oxygen. Most other metals do react with oxygen, but some react faster than others.

Potassium is a metal that reacts with oxygen as soon as a newly cut surface is exposed to air. So much thermal (heat) energy is given out in this reaction that a lump of potassium will burst into flames if left in oxygen. We say that potassium is a very **reactive** metal. The word equation for the reaction is:

potassium + oxygen → potassium oxide

a **Give two pieces of evidence that show this is a chemical reaction.**

Do you remember?

When substances react with oxygen they form compounds called **oxides**.

How other metals react with oxygen

Zinc also reacts with the oxygen in air and will lose its shine after a while. But zinc does not give out as much energy as potassium when it reacts. It can safely be left in the air without bursting into flames. Zinc is less reactive than potassium, but will still burn if heated.

zinc + oxygen → zinc oxide

b **Iron is just a little less reactive than zinc.**
- **(i) What common substance forms when iron reacts with the air?**
- **(ii) What other substance is needed for this to happen?**

Copper is not very reactive. It reacts much more slowly than zinc or iron. Even when heated, copper just forms a layer of black copper oxide that dulls the surface. But this is still more reactive than gold!

copper + oxygen → copper oxide

As in all chemical reactions, the copper and oxygen atoms are rearranged in the new compound.

Zinc burns with a bright blue-white flame.

c **The piece of copper in this photo had a mass of 1 g before it was heated and reacted with oxygen from the air. If the oxide-coated piece was weighed after heating, would its mass have gone down, stayed the same or gone up? Explain your answer.**

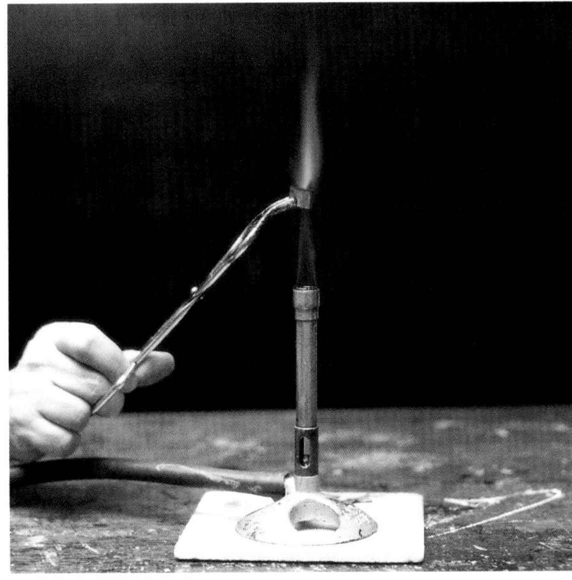

Copper does not burn but it does react to form an oxide layer.

The strange case of silver

Metals can also be tarnished when they react with other gases. Silver (Ag) is another unreactive metal that is used for jewellery. Unlike gold, silver will tarnish and turn black in time. But this is not an oxide layer, as silver does not react with oxygen at low temperatures. Instead it is silver sulphide (Ag_2S). Look at the table and try to work out what is happening.

Where silvery jewellery has been kept	How badly tarnished after one year
in a museum case, in purified air	still shiny
in the laboratory, with hydrogen sulphide gas (H_2S)	completely black
in the country, away from cities	quite shiny
in the city	dull with thin black film
near some large factories	black coating
on a volcanic island far from the mainland	black coating
in the country, near a swamp	black coating

d (i) **What gas seems to be responsible for tarnishing silver?**

(ii) **Write the word equation for the reaction.**

(iii) **What evidence suggests that this problem might be caused by pollution?**

(iv) **What evidence suggests that this problem might also have natural causes?**

Questions

1 List the metals copper, gold, zinc, potassium and iron in order of reactivity, with the most reactive at the top.

2 Malcolm is investigating the reactivity of three metals: potassium, tin and copper. Malcolm's first idea is to heat a small piece of each material in a Bunsen burner flame.

- **a** Tin reacts when heated with oxygen to form tin oxide. Write a word equation for this reaction.
- **b** Why is Malcolm's plan not safe? (Which part would be very dangerous?)
- **c** Malcolm was given a piece of platinum wire to test instead of potassium. Platinum is used in expensive jewellery. What do you think happened when he heated it? Explain your reasoning.

3 Magnesium is a more reactive metal than zinc. It is used in the building of battleships. This is safe under normal conditions, but during the Falklands War a British battleship was struck by an Excocet missile. After the initial explosion a terrible fire started, giving off great heat and a blinding white light, and producing masses of white smoke. What do you think might have happened?

For your notes:

- Metals react with oxygen to make metal **oxides**.
- The more **reactive** the metal, the faster it reacts.
- The more reactive the metal, the more energy is released when it reacts.

F2 Corrosive liquids

Learn about:

- Metals reacting with acid and water

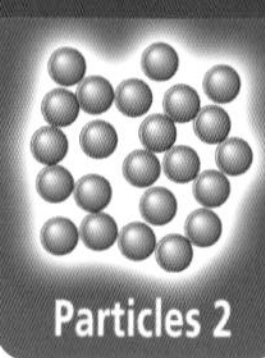

Dangerous acids

Acids such as the dilute sulphuric acid used in car batteries are very strong and corrosive. They will dissolve many metals, producing explosive hydrogen gas and a salt.

iron + sulphuric acid → iron sulphate + hydrogen

This careless mechanic could be making iron sulphate from his spanner!

How different metals react with acid

Hydrochloric acid reacts with metals to form chloride salts. Zinc fizzes steadily in hydrochloric acid as zinc chloride and hydrogen gas are made. The test tube warms up as thermal (heat) energy is released in the reaction.

zinc + hydrochloric acid → zinc chloride + hydrogen

When magnesium reacts with acid, the bubbles of hydrogen are produced more quickly. Magnesium is more reactive than zinc.

Copper is less reactive than zinc. In fact, it is so unreactive that it never reacts with an acid to give hydrogen. But some metals are much more reactive than zinc. Potassium is so reactive that it would be far too dangerous to react it with acid.

Do you remember?

In Unit E Reactions of metals and metal compounds you learnt how metals react with acids.

metal + acid → salt + hydrogen

a **Sue puts magnesium, copper and zinc into hydrochloric acid in separate test tubes. She measures the temperature rise in each. Predict her results.**

Volume of hydrogen produced

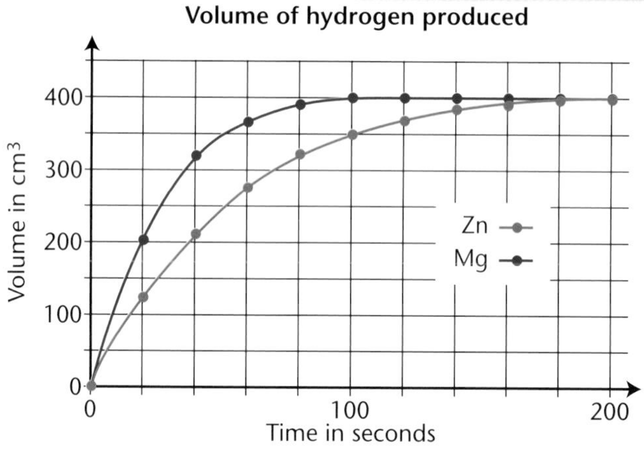

Faster, not more!

Magnesium in acid produces hydrogen faster than the zinc, but will it produce more gas from the same amount of acid? If you collect the gas using a gas syringe you find that, for a given amount of acid, both metals produce the same volume of hydrogen – the magnesium just gets there quicker!

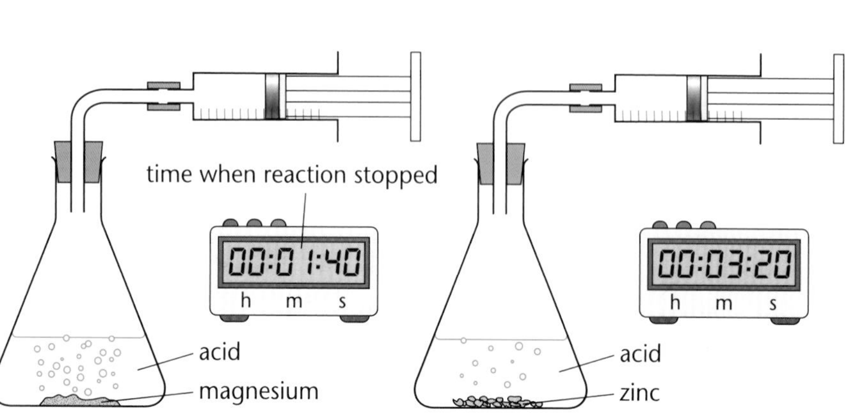

Do you remember?

Aluminium foil stays shiny, so you might think that aluminium is unreactive. In fact, it is reacting to form a very thin but strong oxide layer which seals off the metal and stops further reaction. Even so, aluminium saucepans shouldn't be used for boiling up fruit. The natural acids in fruit can eat through this layer and start a reaction.

Some metals react with water

A small piece of potassium dropped into water reacts in a most spectacular way. The potassium and water react to produce potassium hydroxide and a gas that lights with a squeaky pop. So much energy is given out that the gas catches fire!

potassium + water → potassium hydroxide + hydrogen

b What gas is produced when potassium reacts with water?

Sodium reacts in a similar way, but not quite as violently as potassium. Even so, the sodium gets so hot that it melts and whizzes across the surface of the water on a raft of hydrogen bubbles. Sodium hydroxide is produced.

c Write a word equation for the reaction of sodium with water.

Sodium in water also produces a flame if it is trapped on filter paper.

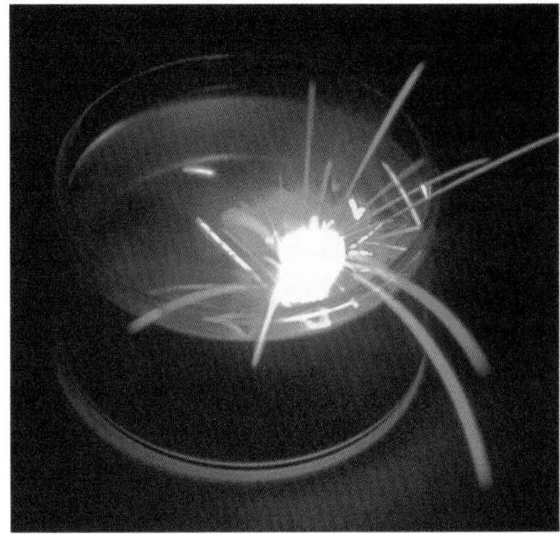

Potassium in water.

Do you remember?

All metal hydroxides are bases. Water-soluble hydroxides such as potassium hydroxide and sodium hydroxide are called alkalis. Because of this, sodium and potassium are called the 'alkali metals'.

Calcium also reacts with water, producing calcium hydroxide and hydrogen. Less energy is produced than when sodium reacts, so it is safe to do the reaction in a test tube rather than an open container.

calcium + water → calcium hydroxide + hydrogen

Most other metals either react very slowly with water or are not affected at all.

Calcium in water.

Questions

1 Sue is investigating the reactivity of metals by putting them in dilute sulphuric acid. She would like to include copper, magnesium and sodium in her investigation.

- **a** Suggest which metal is too dangerous for Sue to use. (She can replace it with zinc.)
- **b** Write a word equation for the reaction between magnesium and sulphuric acid.
- **c** Predict the outcome of Sue's experiment.
  - **(i)** Which metals will react?
  - **(ii)** Which will react most quickly?

2 John wrote in his project on metals that potassium is so reactive it has to be stored under water to stop it from reacting with the air. Explain why that would not be a good idea.

3 Hassan reacted excess magnesium with 100 cm^3 of dilute acid and collected the gas. He had 120 cm^3 of gas when the reaction stopped after 60 seconds. He then repeated the experiment with excess zinc instead of magnesium. How long did it take to collect 120 cm^3 of gas from the zinc and acid? Use the graph on page 54 to help you. (*Hint*: the hydrogen comes from the acid, not the metal.)

For your notes:

- Some metals react with acid to make hydrogen and a salt. The type of salt depends on the type of acid used.
- Very reactive metals react with water to make hydrogen and a metal hydroxide.

F3 Changing places

Learn about:

- Metals displacing other metals

Displacement reactions

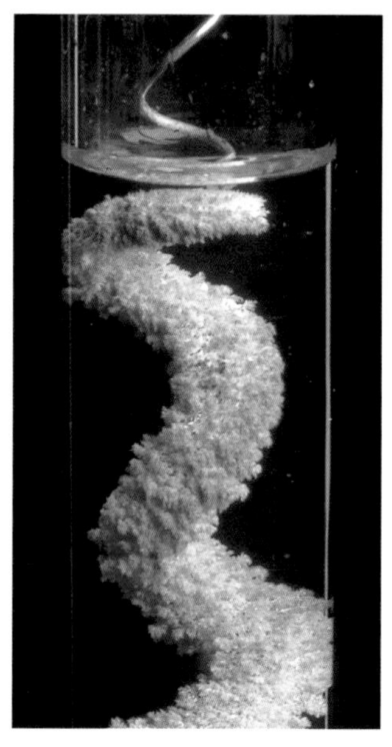

Look at this photo. If you hang a copper wire in a colourless solution of silver nitrate, the solution gradually turns blue, while beautiful crystals of silver metal grow out from the wire.

The copper is more reactive than silver and pushes the silver from its compound. The copper particles take the place of the silver particles, moving them out of the way. This is called a **displacement reaction**.

copper	+	silver nitrate	→	silver	+	copper nitrate
brown solid		colourless solution		silvery solid		blue solution

a **The solution got warmer during this reaction. Why was that?**

Who's most attractive?

There is a simple model you can use to help you understand this type of reaction. Silver and nitrate make a happy couple – until copper comes along. Copper is more 'attractive' than silver, so nitrate pairs up with copper instead. The silver atoms go off on their own.

The more reactive a metal is, the more it 'attractive' it is, and so more likely to pair up with other particles and form a compound. Any metal can be pushed out of its compounds by a more reactive metal.

b **An iron nail gets coated with copper if you dip it into copper sulphate. Explain why this happens. Write a word equation for the reaction.**

Who gets the oxygen?

If you gently heat a mixture of zinc and copper oxide they react together, giving out a lot of energy. Zinc is more reactive than copper so is able to displace it from the compound. In this reaction, the oxygen finds zinc more attractive and copper goes off on its own!

zinc	+	copper oxide	→	zinc oxide	+	copper
Zn	+	CuO	→	ZnO	+	Cu

c **If you heat the zinc oxide and the copper, will you get the zinc back? (If not, why not?)**

Energy changes in displacement reactions

The amount of energy given out in a displacement reaction depends on the difference in reactivity between the metals. You will get a bigger temperature rise if you use magnesium to displace copper from copper oxide than if you use zinc because magnesium is more reactive than zinc.

Using displacement to stop rusting

When iron rusts it reacts with oxygen to make iron oxide. Putting a more reactive metal such as zinc against the iron can stop rusting. The zinc then reacts with the oxygen instead of the iron.

This way of preventing iron rusting is used for ships, submarines and oil rigs. Huge lumps of reactive metal are fixed onto the sides. Oxygen then reacts with this metal instead of the iron.

'Tin' cans for food are made from steel (iron) coated with a layer of less reactive tin. This can last a long time, but if you find a really old tin can out in the open, it will sometimes be just a thin layer of tin over rust!

d **These pupils are talking about what happens to tin cans if they are left outside for a long time. Which explanation do you think is correct?**

e **Sometimes iron is coated with zinc to stop it from rusting. If the zinc is scratched and the iron is exposed it still does not rust. Why is this so different from a tin can?**

Questions

1 Magnesium is more reactive than nickel. Nickel sulphate makes a green solution. Magnesium sulphate makes a colourless solution.

- **a** Write a word equation for the reaction that occurs when a strip of magnesium is placed in nickel sulphate solution.
- **b** What would you observe during this reaction?

2 **a** When a metal, such as zinc, reacts with an acid, the metal displaces the hydrogen. Which is more reactive, zinc or hydrogen?.

- **b** Copper can never displace hydrogen from acid. Which is more reactive, copper or hydrogen?
- **c** Arrange these three metals (zinc, lead and copper) in order of reactivity.
- **d** Where does hydrogen fit into this series? Explain your answer. (*Hint*: which metal can never displace the hydrogen?)

3 Powdered aluminium will react with iron oxide. The reaction is like the one between zinc and copper oxide. Melted iron and aluminium oxide are made in the reaction.

- **a** Which is the more reactive metal, aluminium or iron? Give reasons for your answer.
- **b** Write a word equation for the reaction
- **c** Why is the iron melted, rather than solid?
- **d** Suggest why the reaction works better with powdered aluminium rather than a lump of aluminium.

Do you remember?

Iron rusts when it reacts with air and water.

For your notes:

- More reactive metals push less reactive metals out of their compounds. These reactions are called **displacement reactions**.
- In a displacement reaction, the more reactive metal ends up in the compound.
- Displacement reactions can be useful.
- Displacement reactions often release a lot of energy.

F4 Who's top of the league?

Learn about:

- Making a reactivity series

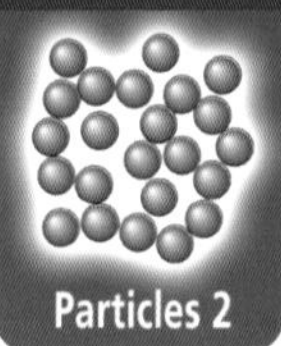

The reactivity series

The **reactivity series** is a list of metals with the most reactive at the top and the least reactive at the bottom. Metals are put into a reactivity series according to how they behave in chemical reactions.

You are going to learn about creating a reactivity series by following Joel's investigation. He made a reactivity series for 10 metals: calcium (Ca), copper (Cu), gold (Au), iron (Fe), lithium (Li), magnesium (Mg), nickel (Ni), zinc (Zn), sodium (Na) and potassium (K).

a **Joel looks at the periodic table on the wall. Potassium and sodium are in Group 1 along with lithium. They have to be stored under oil. Does that mean they are reactive or unreactive?**

Joel's teacher tells him that potassium and sodium are too dangerous for him to use, so he uses the Internet to find out more about them. From his research he discovers how they react with water and records this information.

Reactions with water and dilute acid

Next, Joel tests as many of the remaining metals as he can with water, and then with dilute acid. His teacher tells him he can't use lithium and calcium with acid. He writes up his observations in this table, along with his notes from the Internet.

Metal	Reaction with cold water	Reaction with dilute acid
calcium	many bubbles, made quickly	not tested
copper	no reaction	no reaction
gold	no reaction	no reaction
iron	no reaction	no reaction
lithium	most bubbles, made very quickly	not tested
magnesium	a few bubbles, made very slowly	many bubbles, made very quickly
nickel	no reaction	no reaction
potassium	*Internet notes:* violent reaction – explosive	not tested
sodium	*Internet notes:* vigorous reaction – whizzes about	not tested
zinc	no reaction	some bubbles, made quickly

b **After testing with cold water, Joel lists the metals he can in order of reactivity. Write out his list.**

c **After testing with dilute acid, Joel finds he can make a list of six of the metals in order of reactivity. Write out his list.**

Using displacement reactions

Joel managed to sort out the top six metals into a reactivity series, but he was not sure of the order of the last four: copper (Cu), gold (Au), iron (Fe) and nickel (Ni).

He decides to use displacement reactions to sort out the last four metals. For his experiment he needs a solution of a compound for each metal. He decides to use the sulphates. Unfortunately, Joel cannot get any gold sulphate. But he continues his experiment with the other three: $FeSO_4$, $CuSO_4$ and $NiSO_4$.

The table on the right shows how he sets up the combinations of metals and compounds. Joel records the colours of the solutions at the beginning and the end of the reaction. If there is a colour change, then a reaction has occurred.

Metal compound	Metal: nickel		Metal: copper		Metal: iron	
nickel sulphate			before	after	before	after
copper sulphate	before	after			before	after
iron sulphate	before	after	before	after		

Joel then tries putting some gold metal with each of the sulphate solutions. The gold does not displace any of the other metals from their compounds.

Joel summarises his displacement results as word equations.

copper + nickel sulphate → NO REACTION
iron + nickel sulphate → iron sulphate + nickel
nickel + copper sulphate → nickel sulphate + copper
iron + copper sulphate → iron sulphate + copper
nickel + iron sulphate → NO REACTION
copper + iron sulphate → NO REACTION
gold + any metal sulphate → NO REACTION

Answer questions **d** to **i** to interpret Joel's results.

d **Use Joel's results from the displacement reactions to decide:**
  (i) which metal displaces the other two metals from their compounds
  (ii) which metal is pushed out of its compound by the other metals.

e **Out of nickel, copper and iron, which is:**
  (i) the most reactive?
  (ii) the least reactive?

f **Is gold more or less reactive than the other three metals? Give your reasons.**

g **Write a reactivity series for copper, gold, iron and nickel.**

h **Write a reactivity series for all 10 metals.**

i **Use Joel's reactivity series to predict what will happen when zinc sulphate is mixed with:**
  (i) magnesium
  (ii) iron.

Questions

1 Sam put some copper coins in silver nitrate solution. After some time, the coins went silvery, and the solution started to have a slight green-blue colour.

a Is silver below or above copper in the reactivity series? Give reasons for your answer.

b Write a word equation for the reaction that happened.

c When ancient silver jewellery is dug up, it is usually badly corroded. Is silver above or below gold in the reactivity series?

d Would a gold coin become coated with silver when placed in silver nitrate solution? Give a reason for your answer.

2 Use the reactivity series shown on the right to predict whether reactions will occur when the following substances are mixed. Write word equations for the reactions that do occur.

magnesium *(most reactive)*
aluminium
zinc
copper
silver *(least reactive)*

a silver oxide and magnesium

b zinc and aluminium oxide

c copper oxide and silver

d zinc and silver oxide

3 Rosie has to work out a strategy for finding the reactivity series for these four metals: silver (Ag), sodium (Na), scandium (Sc) and strontium (Sr). Suggest a strategy for Rosie to use. Do not forget to consider safety.

For your notes:

- The **reactivity series** is a list of metals with the most reactive at the top and the least reactive at the bottom.

F5 Reactivity in action

Learn about:

- How the reactivity series can be used

Mending a broken rail

A rail has cracked on the express railway line in open country. Molten iron is needed to fill the gap, but the melting point of iron is very high – over 1500 °C. How can you make a small amount of molten iron so far away from a laboratory?

Spectacular displacement

If aluminium powder is mixed with iron oxide and given a kick-start of energy, a very spectacular displacement reaction occurs. The aluminium pushes the iron out of its compound and takes its place.

aluminium + iron oxide → aluminium oxide + iron

So much energy is given out during this reaction that the iron melts. Look at this photo. If this mixture is reacted between the broken ends of a rail, molten iron fills the gap. When this sets, the rail is as good as new.

a **Aluminium is an expensive metal compared to iron. Why do you think this reaction is not used to make iron in large amounts?**

b **Magnesium is above aluminium in the reactivity series. Do you think it could be used instead of aluminium in this reaction? Explain your answer.**

c **If magnesium was used, do you think it would give out more, less or the same amount of energy as aluminium? Explain your answer.**

Bottom is best ...

Corrugated iron is cheap, strong and waterproof, but iron is quite high in the reactivity series. Over 20 or 30 years it will rust away.

Copper is much less reactive than iron and does not react with water or acid. Copper is more expensive than iron, but copper roofs last much longer. Copper's low position in the reactivity series also makes it useful for water pipes and tanks.

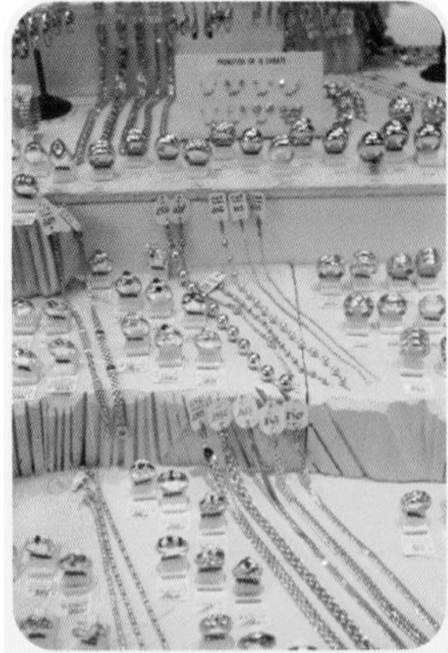

And don't forget that the metals at the bottom of the reactivity league such as gold and silver stay the shiniest!

... unless you're going up

Aluminium is very reactive. The space shuttle's booster rockets burn aluminium as a fuel. Only a very reactive metal will produce a fast enough reaction for this. The white smoke in the photo is aluminium oxide.

Do you remember?

Metals are found in rocks. Some like gold and silver are found as pure metal. Others are found as compounds joined with other elements and are obtained using chemical reactions. A rock containing a metal or a metal compound is called an **ore**.

Reactivity and discovery

The two most common metals found on the surface of the Earth are aluminium and iron. Iron was discovered a few thousand years ago, but aluminium was discovered just a couple of hundred years ago. Why such a difference? And why were the much rarer metals copper and lead discovered thousands of years before iron?

The link is reactivity. The more reactive the metal, the more tightly it is combined in its compounds. You need more energy to break apart the atoms in the compounds, so it is more difficult to extract the metal from its ore. The less reactive the metal, the easier it is to extract.

Copper is not very reactive so the metal could have been made by accident in a fire, when copper ore reacted with carbon from the wood. Iron is more reactive. Iron ore reacts with carbon but needs a much higher temperature. Aluminium ore does not react with carbon at all. It was discoverd much later as it can only be made using electricity.

d **Silver is not very reactive. When do you think it was discovered, compared to iron and aluminium?**

Questions

1 Lead is a poisonous metal that is just above copper in the reactivity series. It can be pressed into thin, watertight sheets that can be easily moulded into shape.
   - **a** What properties make lead useful as a roofing material?
   - **b** Water pipes used to be made of lead. Suggest why this is no longer the case, especially in areas where the water is slightly acidic.

2 Most metals are found combined with oxygen in the rocks. The metal must be got from these ores before we can use it. Metals such as iron, copper, zinc and lead are displaced from their ores using carbon (in the form of coal).
   - **a** Look at a copy of the periodic table. Is carbon a metal or a non-metal?
   - **b** Where must carbon be in the reactivity series compared to iron, copper, zinc and lead?
   - **c** What compound forms when carbon reacts with lead oxide?
   - **d** Write the word equation for this reaction.
   - **e** Aluminium cannot be got from aluminium oxide by this reaction. Where must carbon be in the reactivity series compared to aluminium?
   - **f** Aluminium is the commonest metal yet it was discovered only relatively recently. Suggest a possible reason for this.

For your notes:

- The **reactivity series** can be very useful for extracting metals from their compounds.
- The uses of different metals depend on their reactivity.

F6 Variables together

Think about:
- Variables that interact

Iron ... rusts away!

Iron is the second most common metal in the Earth's crust. More iron is produced and used than all of the other metals put together. Iron is very strong. It is used to make cars, bridges, the skeletons of buildings, ships, nails and screws, steel cables, pans, radiators, bicycles and thousands of other items.

But there's a problem. Iron rusts! Iron reacts with water and the oxygen in air, making flaky brown iron oxide. This breaks off, exposing fresh metal. So rusting can carry on until there is no metal left. Preventing rusting saves billions of pounds of repairs to iron structures.

How can we stop iron rusting?

Before we can stop iron rusting, we have to understand the variables that affect the way iron rusts.

a **Explain how each of the following affects rusting:**
(i) covering a surface with paint
(ii) oiling a surface
(iii) making an object out of aluminium
(iv) bolting a lump of aluminium onto iron
(v) being in a desert
(vi) being at the North Pole
(vii) being at the seaside.

I'm going to put one nail in a tube of oil so that no oxygen or water can get in.

I'll put another nail in a tube of air. Oxygen will be there but I'll put some silica gel in too, to make sure the air is dry.

I think that the nail in the air with oxygen will rust.

I will put a nail in water, but I will boil the water first to drive off any oxygen. I can compare this to your first tube that has no water or oxygen.

I think my nail in water will go rusty.

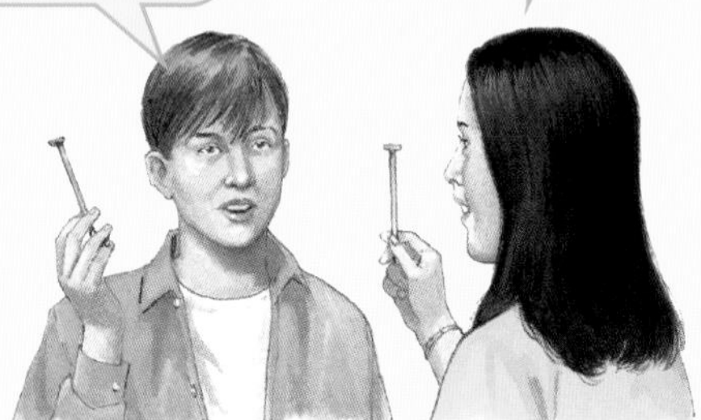

Planning an experiment

Terry and Tasha decide to do an experiment to see which of their chosen variables controls the way iron rusts.

All three tubes were kept at the same temperature to control that variable.

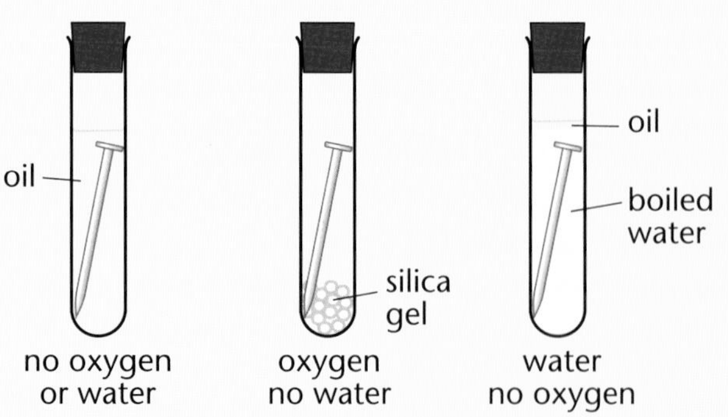

One week later they checked their nails. None of the nails had gone rusty!

Analysing the results

This experiment shows that oxygen and water on their own will not make iron rust. Terry and Tasha decided to set up another nail in a tube - this time just half full of plain water, so that both oxygen and water are present. They left it a week at the same temperature, and found that this time the nail started to rust.

	No oxygen	Oxygen present
no water	no rust	no rust
water present	no rust	rust present

Terry and Tasha showed that both oxygen and water are needed to make iron rust. Rusting only occurs when the two variables **interact**.

Going further

Lennox had heard that cars rust faster in the summer than in the winter. This surprised him, as he would have thought it would be the other way around. After all, it rains more in winter. Petra reminded him that chemical reactions often speed up if they are heated. It is certainly warmer in the summer.

They decided to design an experiment to see if temperature affected the speed at which iron rusted. Petra predicted that the iron would rust faster if the temperature increased.

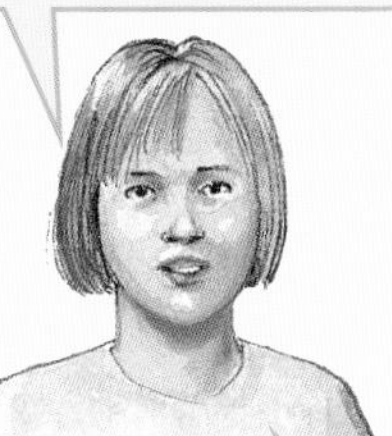

Temperature	Wed	Thur	Fri	Mon	Tue	Wed	Thur	Fri	Mon
fridge 10 °C	✗	✗	✗	✗	✗	✗	✗	✗	✔
bench 25 °C	✗	✗	✗	✗	✗	✔	✔	✔	✔
water bath 40 °C	✗	✗	✗	✔	✔	✔	✔	✔	✔
water bath 55 °C	✗	✗	✗	✔	✔	✔	✔	✔	✔
water bath 70 °C	✗	✗	✔	✔	✔	✔	✔	✔	✔
water bath 85 °C	✗	✔	✔	✔	✔	✔	✔	✔	✔

b (i) **Was Petra's prediction correct? Write a conclusion for her experiment.**

(ii) **The nails at 40 °C and 55 °C seem to have rusted after the same time. Can you see why that result might not be very reliable?**

(iii) **How easy do you think it is to decide when the nail first goes rusty?**

(iv) **Suggest one or two ways in which you could improve the design of this experiment.**

(v) **Terry thought he'd try another test tube at –5 °C in the freezer. What do you think would happen?**

Questions

1 Salt water speeds up rusting.

a Salt is put on the roads to stop them icing up in winter. Why do you think it is a good idea to wash the underside of a car after the winter?

b Design an experiment to see how the amount of salt in water affects the speed of rusting.

2 Imagine that you are a scientist working for a company making bolts that are used to build bridges and buildings. Your job is to test all the new bolts to see how rustproof they are. Rusting is a slow process, so you need to set up conditions which will make the bolts rust as quickly as possible. Design a 'rusting box' for maximum rusting.

3 Emma liked the way the copper roof on her local church had turned a beautiful green colour and wondered what caused this. She thought that water and oxygen might be involved as they caused iron to rust. She also knew that the green compound was copper carbonate, so perhaps carbon dioxide was needed too. Design an experiment for her to investigate which, if any, of these three substances are needed to turn copper green.

G1 Environmental chemistry

Learn about:

- The soil
- Pollution

It's all around us

The land, air and sea make up our environment.

We live on the land. The rocks beneath us form the soil in which we grow our food.

The air we breathe is a mixture of gases. We'd soon die without the oxygen it contains.

Rivers carry rain water down to the sea. We need fresh water to drink.

Rocks and soil

You can't grow crops in rock. Plants need soil to grow. But most of the nutrients that plants need come from the minerals in rock. Soil forms when the rocks are physically and chemically weathered over tens or even hundreds of years. Chemical compounds in the minerals react with water and air, producing nutrients that dissolve in water for plant roots to take up.

Soils made from granite contain a lot of sand and clay. Many plants, such as heather, thrive in this kind of soil and crops grow well in it. These soils can be naturally slightly acidic (pH 5–6). This acidity does not come from the damaging effects of acid rain pollution (see page 66).

Soils made from limestone or chalk have less sand and clay. They are often in thin layers, do not contain as many nutrients, and dry out very easily. Plants do not grow so well. These soils are naturally alkaline (pH 8–10). Some plants such as 'old man's beard' thrive on this, but heathers don't.

Do you remember?

Weathering is when rocks break down. In physical weathering, the rocks are simply broken into smaller pieces. In chemical weathering, chemical reactions make new substances.

Scottish heather.

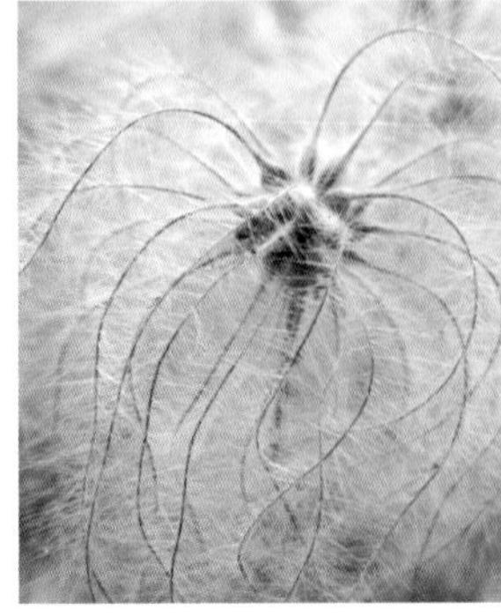

Old man's beard (wild clematis).

a **Heathers and azaleas thrive in acidic soil. Clematis and camellias prefer alkaline soil. Suggest the best plants for these gardeners.**

(i) Andy lives in Aberdeen (Scotland). His garden soil is pH 6.

(ii) Henrietta lives in Chiselhurst (Kent). Her garden soil is pH 9.

You can test soil to see if it is acidic or alkaline by shaking it with a little water and some universal indicator. Some flowers acts as indicators, too. Hydrangeas have pink flowers when grown in alkaline soil, but blue flowers in acidic soil.

b **Fiona bought a beautiful blue hydrangea in Cornwall, but when she put it in her garden in London it turned pink. What happened?**

Looking after the soil

Soil is precious! When building new houses, soil used to be just dumped and garden areas were filled with rubble at the end. These days all 'topsoil' is carefully removed before building work starts. It is stored and put back in place after building work is finished.

But pollution can damage soil. Many old factory sites have been cleared for building homes. If the soil contains factory waste, such as poisonous metal compounds, plants will not grow well. It may also be a health hazard for humans.

Looking after the air

About 50 years ago, most people in London heated their homes with open coal fires. The smoke and soot went up the chimney and got caught in the autumn fog to form a thick black smog – known as a peasouper.

Eventually Parliament passed the Clean Air Act and banned coal fires, which soon cured the problem in London. People then had to use 'smokeless' coke for their fires. This was made by heating coal in special ovens and driving off the most polluting gases. Coke was made in South Wales, which unfortunately then became highly polluted instead of London.

These days our city air pollution is caused by exhaust fumes from large numbers of cars and lorries.

Looking after the water

We use water for drinking, washing, cooking and cleaning. We eat fish from the sea and enjoy our holidays at the seaside. But we also sometimes pollute our water resources. Around 150 years ago, all the sewers of London flowed into the River Thames. In the summer, the MPs in the Houses of Parliament had to put up with the 'great stench' from the polluted river. Even 50 years ago the Thames was so polluted that you had to go to hospital and have your stomach pumped out if you fell in.

Today, sewage is carefully treated before the water goes back into rivers. Scientists like the one in this photo take samples and monitor the pollution levels. If a company pollutes a river, they can be heavily fined by law. Thanks to improved technology and laws, fish have returned to the River Thames. Some people even swim in it!

Questions

1. The Cairngorm Mountains of Scotland are made of granite. A scientist monitoring the soil on the mountains found that it was quite acidic. Should he look for a local source of pollution? Explain your answer.
2. Cars have been improved so they cause much less pollution than they used to. Why is air pollution from cars still such a big problem in our cities?
3. Suggest two ways in which we know that the River Thames is less polluted than it used to be.
4. Explain why the Clean Air Act was only a *local* solution to the problem of air pollution.

For your notes:

- Soils form slowly as rocks are weathered physically and chemically.
- Soil is needed to grow plants.
- The land, air and sea can all suffer from **pollution**.
- Science can help to reduce pollution problems.

G2 Acid rain

Learn about:

- Acid rain and its sources

What is acid rain?

All rain is naturally very slightly acidic because of the dissolved carbon dioxide in it. This is a very weak acid with a pH of about 6. Over hundreds of years, all rain will chemically weather limestone rocks, statues and buildings.

carbon dioxide + water → carbonic acid
$CO_2 + H_2O \rightarrow H_2CO_3$

But other gases dissolve in rain to make it much more acidic – as acidic as the vinegar you put on your chips! This **acid rain** is about pH 3–4.

a **What problems do you think this acid rain might cause?**

What causes the problem?

Sulphur is a yellow solid that burns in air to give an acidic gas called sulphur dioxide.

sulphur + oxygen → sulphur dioxide
$S + O_2 \rightarrow SO_2$

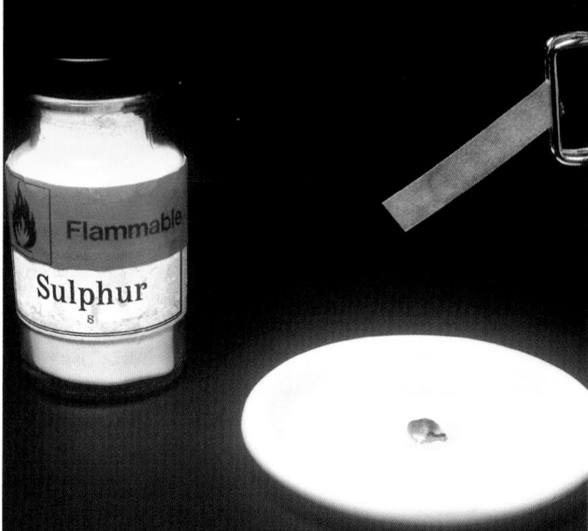

Burning sulphur makes an acidic gas.

If this gas gets into the air, it reacts with more oxygen and rain water to form a weak solution of sulphuric acid. This is a stronger acid with a pH as low as 4.

Where does it come from?

There is sulphur in volcanoes. Every time a volcano erupts, millions of tonnes of sulphur dioxide are sent up into the air. This causes a lot of acid rain.

Lots more sulphur dioxide gets into the atmosphere during forest fires, or from bubbling swamps. Plants contain some sulphur and this is released as sulphur dioxide when they burn. Plants also rot in swamps, and produce hydrogen sulphide gas, which turns to sulphur dioxide in time.

Overall, about half of the acid rain has 'natural causes' – but there's not a lot we can do about these natural forms of air pollution!

Where do we come in?

There is also a lot of sulphur in fossil fuels such as coal, oil and gas. When we burn fossil fuels, the sulphur reacts with oxygen to produce sulphur dioxide. If this gets into the air it produces more acid rain. We burn millions of tonnes of our fossil fuels every day, so we make the problem of acid rain worse.

Don't forget that the electricity you use is produced in power stations. Many power stations burn fossil fuels, and these can produce millions of tonnes of sulphur dioxide over the years. Every time you use electricity, you are causing a little more air pollution.

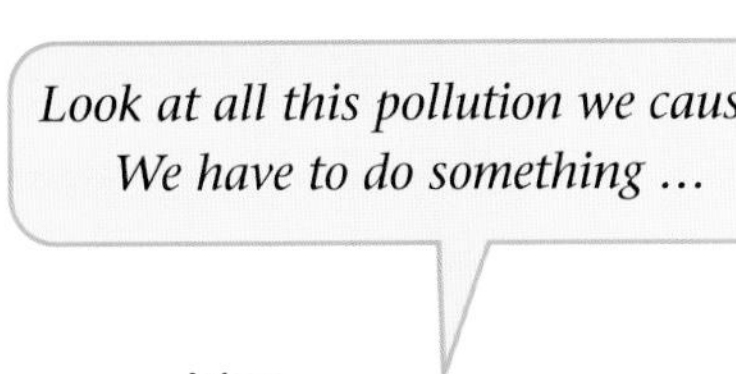

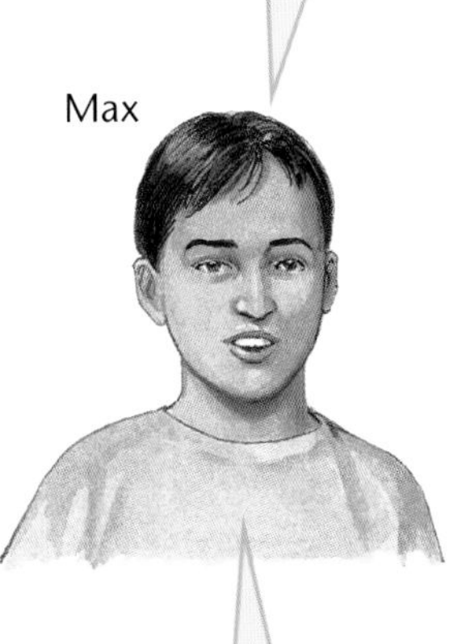

Natural sources – volcanoes

Man-made sources – burning fossil fuels

Global sources of acid rain.

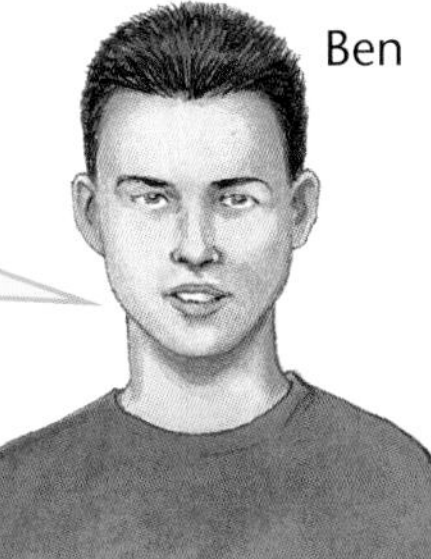

But look – as much acid rain is caused by volcanoes! What can we do about that?

True, but we should still try to stop the pollution we do cause ...

b Who do you agree with, and why?

Nasty cars ...

Cars produce sulphur dioxide, but their exhaust gases also contain lots of other nasty chemicals. Poisonous carbon monoxide and choking nitrogen oxides mix with tiny particles of sooty carbon. This makes walking or cycling along a busy city road an unpleasant experience. When it is hot and sunny, they react with oxygen in the air to make poisonous ozone. Cities like London are often blanketed by a hazy brown smog in the summer.

Questions

1 Nature may make as much sulphur dioxide as we do in the world as a whole, but locally the balance may be different. Suggest a local source for each of these acid rain problems.
   - **a** Many factory towns in England were badly affected by acid rain in the 1950s.
   - **b** The pine forests of Germany have been damaged where busy motorways pass through them.

2 **a** Name one source of carbon dioxide in the air.
   - **b** Another 'acid gas' can form naturally during thunderstorms. What do you think it could be? (*Hint*: lightning makes the two commonest gases in the air react together.)

3 Maria lives on the Italian Island of Vulcano. There are no large power stations or factories and few cars on the island. She says the air is pure and pollution-free. Is she correct? Explain your answer

4 Petrol explodes with air inside a car engine. The exhaust gases include carbon dioxide (CO_2), water vapour (H_2O) and traces of nitrogen oxides (NO and NO_2).
   - **a** What reaction produces the carbon dioxide and water? Write a word equation for it.
   - **b** What reaction do you think produces the nitrogen oxides?
   - **c** What do you think happens to the nitrogen oxides when they get into the air and react with water?

For your notes:

- Rain water is always slightly acidic.
- **Acid rain** is much stronger. It forms when sulphur dioxide pollutes the air.
- Sulphur dioxide comes from natural sources such as volcanoes, and from burning fossil fuels.

G3 More about acid rain

Learn about:
- The effects of acid rain
- How to prevent and cure acid rain

So what's the problem?

Acid rain is very corrosive. It makes iron bridges and machinery rust away even faster than usual. It also weathers limestone, which is a natural form of calcium carbonate. Many important buildings are built from limestone, such as Canterbury Cathedral. Acid rain has weakened them. Many wonderful sculptures and statues have also been ruined by acid rain.

a Why do you think that detailed sculptures seem to be damaged much more than solid blocks of limestone?

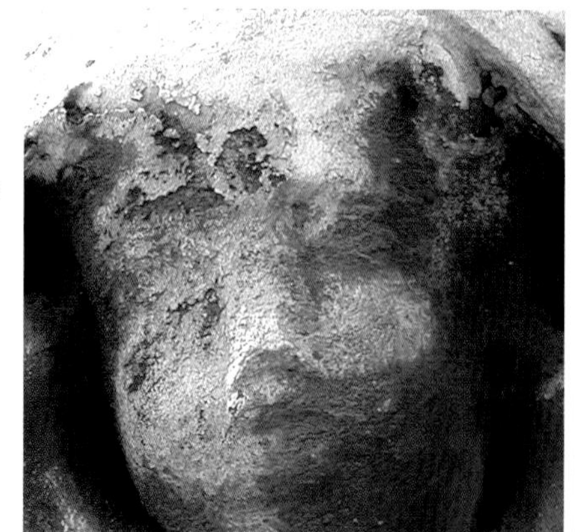
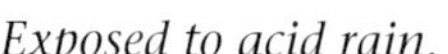
Exposed to acid rain.

Protected from acid rain.

Is that all?

Air pollution and acid rain affect human health and the environment.

- Short-term exposure to high levels of sulphur dioxide may cause coughing, tightening of the chest and irritation of the lungs.
- Acid rain damages the leaves of trees, where food is made. It also damages the bark, making it easier for insects and microorganisms to get in. It washes some essential nutrients out of the soil and releases poisonous aluminium, which can damage plant roots.
- Acid rain poisons fish in lakes. If the pH falls below 5.5 neither the fish nor the organisms they feed on can survive.

Not just a local problem

Power stations burning fossil fuels used to release millions of tonnes of sulphur dioxide straight into the air. The local area was badly affected by acid rain, so tall chimneys were added to carry away the gases in the wind. But the acid rain just fell somewhere else. Much of Britain's acid rain was blown to Norway.

In the past, people just looked for local solutions to air pollution problems. Today we understand that that the problem is global. We have to work to reduce the pollution we cause, not just move it out of our area.

How can scientists help?

Some people blame scientists for the problems of pollution, but in fact scientists help solve the problems. The Government has set limits for the amount of sulphur dioxide in the air. Scientists set up automatic datalogging stations all over the country, **monitoring** the amount of sulphur dioxide in the air. You can log on to these through the Internet and find out just how polluted your area is today.

b 100 parts per billion is considered a 'medium risk' health hazard for sulphur dioxide. Describe the pollution in Bexley over the year shown. Was there a problem?

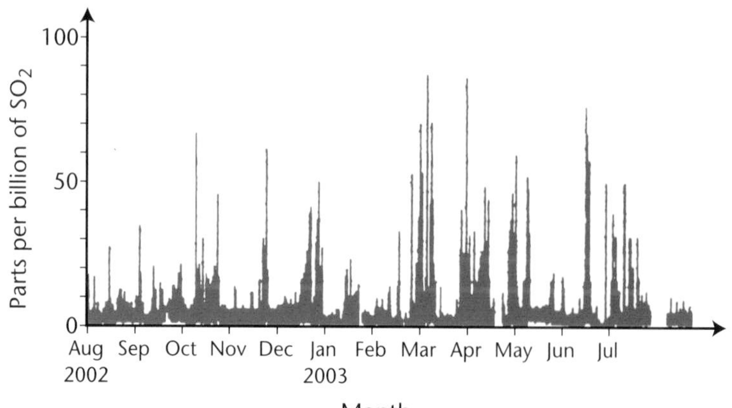

Sulphur dioxide levels in Bexley, London, 2002–3 from a roadside monitoring station.

Other environmental scientists regularly monitor the acidity of rivers and lakes using pH meters. The idea of monitoring is to identify areas of bad pollution so that we can work to reduce it. Here are some cures scientists have come up with.

- Some Scottish lochs lost their fish because of acid rain. Scientists have neutralised the acid waters by adding lime (a soluble base, or alkali). New fish have now been put back into these lochs.

- Scientists helped draw up environmental laws to make coal-fired power stations clean up their waste gases before they are released into the air. They developed special 'gas scrubbers' which wash sulphur dioxide out of the waste gases using an alkaline solution.

- Cars do not produce quite as much air pollution as they used to because most garages now sell 'low sulphur' petrol and diesel.

- Scientists have also helped to reduce the pollution from the other nasty exhaust gases by developing **catalytic converters** for cars. These turn carbon monoxide and nitrogen oxides back into carbon dioxide and nitrogen. All new cars must now, by law, be fitted with catalytic converters.

How can you help?

The more energy you use, the more pollution is produced. We can all help to reduce pollution by changing the way we live. Use a little less energy at home by turning off lights or turning down the heating. Walk, cycle or take the bus instead of going by car!

Questions

1 Write a word equation for the reaction of calcium carbonate with the sulphuric acid in acid rain.

2 The lime used to neutralise sulphuric acid in polluted lakes is calcium hydroxide. Write a word equation for this reaction.

3 The rock beneath some lakes is limestone. Explain carefully why these lakes do not seem to suffer from the effects of acid rain.

4 You live near a lake that has been damaged by acid rain. A nearby quarry owner has offered you a lorry-load of rock chippings that he says will help cure the problem. Design an experiment to see if the rock is suitable for the task.

5 Write an article for a school newspaper suggesting ways in which everyone might help to reduce air pollution by changing the way we live.

For your notes:

- Acid rain weathers limestone buildings, corrodes iron, and kills fish and trees.
- Scientists now **monitor** pollution so they can do something if there is a problem.
- Power stations must now clean their waste gases using an alkaline solution.
- Lime is used to neutralise the effects of acid rain in lakes.
- Petrol and diesel now have a lower sulphur content, and **catalytic converters** remove many polluting gases from car exhausts.
- We can all reduce pollution by making small changes to our lifestyles.

G4 Global warming

Learn about:

- The greenhouse effect
- Global warming

The trouble with burning fuels

Cars produce a nasty mixture of chemicals in their exhausts. Petrol also produces carbon dioxide when it burns. Although this is not poisonous, scientists think that it can cause problems on a global scale.

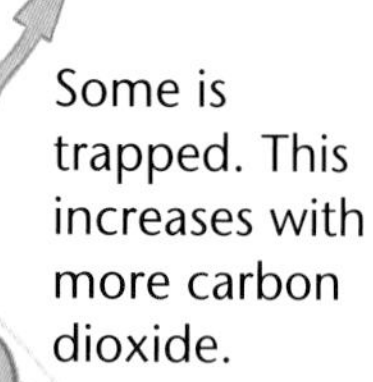

The greenhouse effect

Every day energy reaches the Earth from the Sun. But the Earth also radiates energy into space, so the Earth doesn't get too hot. The energy lost and the energy gained balance to give the Earth a stable temperature. Fortunately for us, this happens to be suitable for living things. This balance is controlled by carbon dioxide in the air, which traps some of the heat energy. This is called the greenhouse effect. Without this natural 'blanket' to keep us warm, the Earth would freeze!

All fossil fuels produce carbon dioxide when we burn them. We have burned so many of our fossil fuel reserves over the last 200 years that the amount of carbon dioxide in the air has gone up by a third! This extra carbon dioxide traps more energy, so the balance changes and the Earth gets warmer. This is called global warming. If it gets out of hand it could cause serious problems for the environment.

a **There is 0.036% of carbon dioxide in the air today. Calculate what it was 200 years ago.**

b **What do you think might happen to the temperature of the Earth if there was a lot less carbon dioxide in the air? Make a list of the problems this might cause.**

Carbon dioxide in the air causes global warming! It traps the Sun's energy and makes the Earth warm up. It's a disaster!

How can carbon dioxide be a problem? We breathe it out, and plants need it to make their food. It's perfectly natural and can't possibly do any harm.

Yes, some carbon dioxide is fine, but too much upsets the balance of gases in the air. We're burning our fossil fuels so fast that we're upsetting that balance!

Does it matter?

Living in Britain, you might not think it would be much of a problem if the Earth warmed up a bit. But things are not that simple. Making the Earth hotter would change the environment.

Some of the ice at the North and South Poles will melt, so the size of the oceans will increase. The sea level will rise and some land will be flooded.

With more energy in the air, the weather will become more violent and unpredictable. There will be high winds and heavy rains in some areas.

In some areas it will get hotter and the rains will dry up. Crops will die as the land turns to desert, and millions of people could starve.

c Who do you think is more likely to suffer from the effects of global warming, people in the developed or developing world? Explain your answer.

What can we do?

The more energy resources we use, the more pollution we make. It's almost as simple as that. So anything we do to save energy can help.

Leaders of the world's developed countries, including Britain, met in 1997 at the Kyoto Summit in Japan. They agreed to cut the amount of carbon dioxide their countries produce by 6%. Unfortunately, most governments are not doing this yet, and the USA has not made a start!

d Look at the table. How significant do you think a 6% cut in Britain's output of carbon dioxide would be? (*Hint*: compare this to the *total* amount produced by all three countries.)

Country	Population	Tonnes of CO_2 produced per person per year
Britain	60 million	9.5
USA	280 million	19.5
India	1000 million	1.1

Questions

1 Air pollution in Athens became so bad that half the cars were banned from driving into the city on any one day.
- **a** How would this decrease air pollution?
- **b** Do you think it was a good idea?
- **c** What other measures can be used to reduce the amount of air pollution produced by cars and lorries?

2 Imagine living in a low-lying coastal area of Britain in the year 2050. Global warming has increased the average temperature by 3°C. The sea level has risen. The annual rainfall has decreased. How would life have changed for:
- **a** a fisherman?
- **b** a farmer?
- **c** a family owning a hotel on the coast road?

3 Britain is as far north as Newfoundland. It is kept surprisingly warm by an ocean current called the Gulf Stream, which flows up from the tropics. Global warming will upset the weather but will also change ocean currents. What might be a possible effect of global warming for Britain? (*Hint*: the *Titanic* sank off Newfoundland.)

For your notes:

- Car exhaust contains a mixture of polluting gases.
- Burning fossil fuels makes carbon dioxide.
- More carbon dioxide in the air increases the greenhouse effect and may cause **global warming**.

Looking at the evidence

Think about:

- The strength of evidence

More about global warming

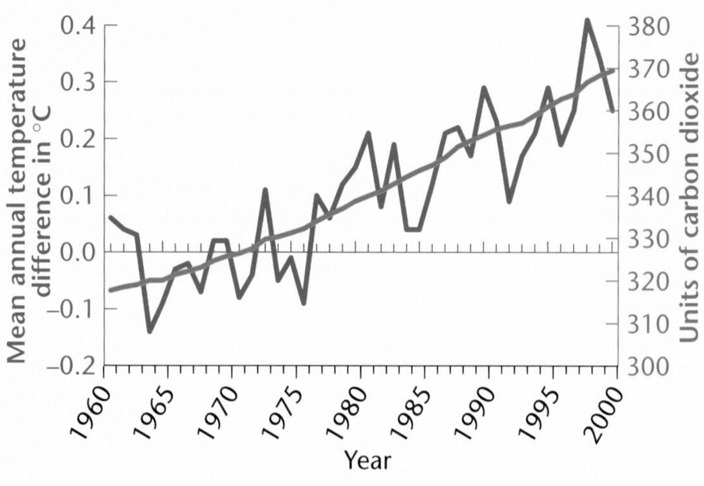

Henry and Leena are studying global warming. They want to look at evidence for the link between carbon dioxide in the air and the temperature of the Earth.

They found some data on the Internet that gave the percentage of carbon dioxide in the air and how the global temperature has changed over the 40 years up to the year 2000. They decided to plot graphs of these figures and compare them, to see if there was a pattern. Here are the graphs they made from the data.

At first glance, both carbon dioxide levels and the mean temperature seemed to have risen. But the pattern is not simple; the temperature graph shows a lot of variability from one year to the next. The temperature in the early 1960s also seemed to have dipped lower for a while, which looked like an **anomaly**.

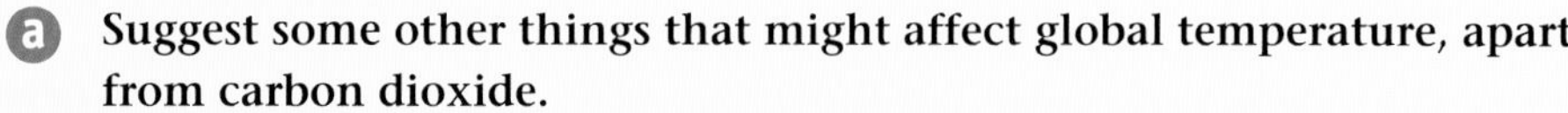

a **Suggest some other things that might affect global temperature, apart from carbon dioxide.**

b **In what way is the temperature data for the early 1960s anomalous?**

c **Even if the two graphs match exactly, would it prove that carbon dioxide causes the global warming?**

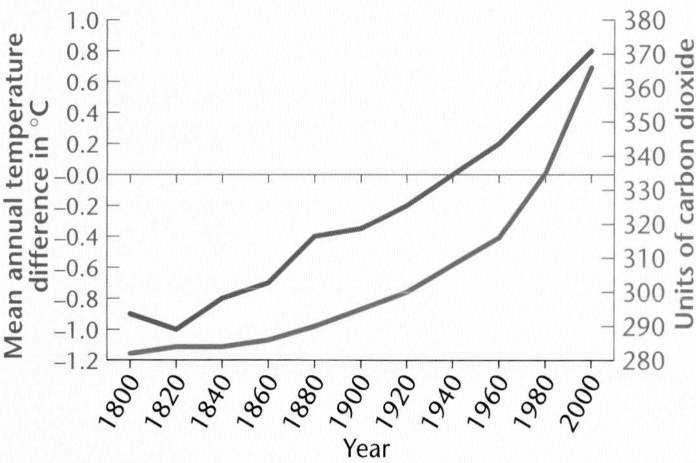

Going further back in time...

Henry and Leena decide that they need more evidence. They want to know if the warmer period in the early 1960s was just an anomaly or part of a wider pattern. They find this graph on the left that looks back 200 years. We have been burning fossil fuels for the last 200 years.

d **What does this graph show about the link between carbon dioxide and temperature over the last 200 years?**

e **This graph shows a similar trend to the one Henry and Leena plotted. What makes the evidence stronger for a link between temperature and carbon dioxide?**

Henry finds another graph showing just temperature for the last 1000 years.

f **Scientists think that carbon dioxide levels during the Middle Ages and Elizabethan Age stayed fairly constant. Does that fit the pattern of the last 200 years?**

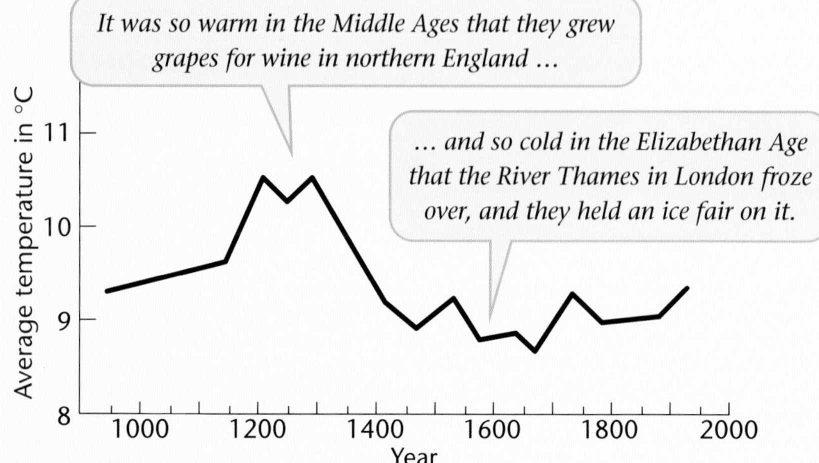

And don't forget the ice ages ...

Around 20 000 years ago, Britain was covered by ice and woolly mammoths roamed the country. So much water was frozen in icebergs that the sea level fell and what is now the English Channel was dry land.

Leena found another graph that showed data going back 250 000 years. Scientists have found a way to work out temperature and carbon dioxide levels from layers of ice in the Antarctic. The ice there has layer upon layer stretching back for a million years or so. The scientists drilled down through the ice, taking samples as they reached older and older layers.

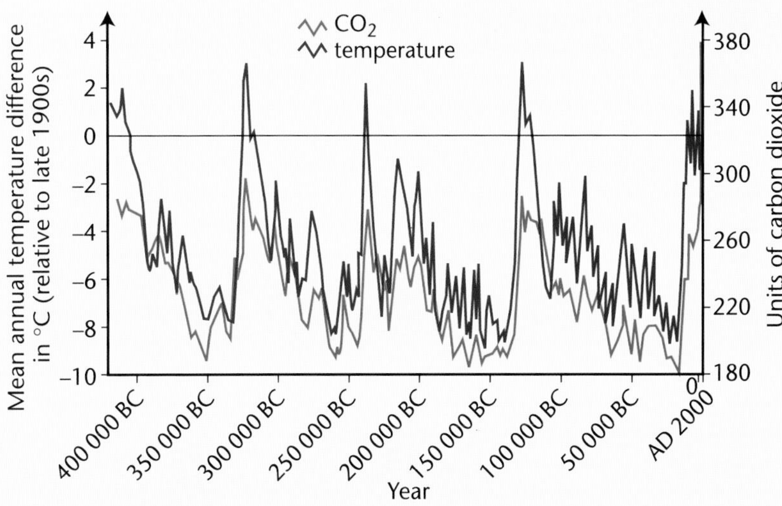

g **Describe the temperature pattern shown by the graph above. Which parts show ice ages?**

h **Look carefully at the variation in carbon dioxide level. Does it seem to change before or after the temperature? Does that support the idea that carbon dioxide is controlling the temperature change or not?**

i **The carbon dioxide level rises dramatically at the end of the graph. Why is that? As this change is so unusual, is it easy to predict what will happen next?**

The graph shows that the Earth's climate has been changing for hundreds of thousands of years. There have been long, very cold ice ages followed by sudden rises in temperature. The carbon dioxide level has been varying naturally without our intervention, but has it been driving climate change or simply following it?

Leena and Henry realise that the Earth's climate is much more complicated than they thought. It could warm up, melt the ice and flood the land – or we could have another ice age. The one thing we can be certain of is that it is unlikely to stay constant. The role of carbon dioxide is not clear. But we have upset the balance by burning fossil fuels and driven the levels to new highs, which is probably not a good idea.

Questions

1 **a** Plot your own line graphs from these figures for the last 200 years.

Year	1800	1820	1840	1860	1880	1900	1920	1940	1960	1980	2000
Carbon dioxide level in parts per million	282	284	285	286	290	296	302	310	321	340	370
Mean average temperature difference in °C	−0.9	−0.95	−0.8	−0.7	−0.45	−0.4	−0.2	0	0.05	0.3	0.7

b Describe the pattern each graph shows and compare the two trends.

c Now plot a scatter graph of carbon dioxide (*x*-axis) against temperature (*y*-axis). Draw a line of best fit through your points and describe any trend shown, including any anomalies.

2 An environmentalist group is campaigning to reduce greenhouse gas emissions from power stations. Their posters say they want to stop climate change. Write an article for your school magazine explaining why you support their campaign but doubt that they'll be able to achieve their stated aim.

H1 Products from reactions

Learn about:
- Useful chemical reactions

Particles 2

Some useful products

Some chemical reactions make useful products. Nearly everything we use has been made from **raw materials** changed by chemical reactions into something more useful. Think about the materials used in a car.

The steel body is a form of iron made from iron ore (iron oxide) by heating it with coke. The molten iron is then converted to steel by dissolving small amounts of carbon in it.

The aluminium in the engine is made from bauxite (aluminium oxide). Bauxite is heated until it melts and the aluminium is split from the compound using electricity.

The rubber for the tyres can be made from the sap of the rubber tree. The sticky sap reacts with sulphur to make it harder and springier.

The plastic for the dashboard is made from chemicals in oil. Crude oil is distilled to separate the chemicals in it. Some of these are then broken up and rearranged by chemical reactions to produce plastics such as polystyrene and PVC.

The glass for the windscreen is made from sand. Pure sand is cleaned with acid before being melted with other chemicals.

a **Draw a flow chart for the processes above, ending with the production of a car.**

The chemistry of food

When you cook food, chemical reactions take place. They break down the food chemicals to make it softer, or easier to digest. They also make it taste better. When you make toast, some of the starch in the bread is broken down into sugar, which makes toast taste sweet. But burnt toast is just carbon!

Proteins in food are also broken down by chemical reactions during digestion, to produce smaller molecules called amino acids. Other chemical reactions then produce new proteins such as those you need for muscle growth and repair.

When food goes bad, the chemical reactions are not so useful. Bacteria and fungi grow on the food, and their chemical reactions produce poisonous waste products. If you eat food that has 'gone off', these poisons can make you very ill.

But sometimes we can use fungi to make special food for us, by using the chemical reactions that take place inside fungi. Read this case study from Dr Vega.

I am a food scientist. I do quality control testing on food. Our raw materials are a microscopic fungus and cheap starch made from potatoes. The fungus feeds on the starch and grows really quickly. Its body is made up of highly nutritious protein. This fungal protein is called Quorn.

Quorn is a really cheap protein that can be used in many different foods. If you use the right chemical reactions, you can make it look and taste like chicken and other meats.

It was first made in the 1960s. Our company had to do tests on the fungi for years to be sure that they were safe to eat. I make sure that every batch of Quorn produced is of good quality.

b Draw a flow chart to show how meat-free 'chicken nuggets' could be made from potatoes.

The chemistry of life

All life relies on chemical reactions.

- Plants make the food they need using a chemical reaction. They use sunlight to combine carbon dioxide and water to make glucose. Oxygen is a waste product.
- All living things respire to get the energy they need for carrying out life processes. In respiration, glucose combines with the oxygen you breathe. Carbon dioxide and water are waste products. This chemical reaction gives out energy and is similar to burning a fuel (see page 76).
- Chemical reactions during digestion make glucose from starchy foods such as bread. The glucose is used for respiration.

c You breathe in oxygen but breathe out carbon dioxide. Where does the carbon dioxide come from?

Questions

1. Copper is used for electrical wiring. The raw material for this is copper ore (copper oxide). Suggest a chemical reaction that could be used to get the copper from this ore.
2. When plants make their own food using sunlight, the chemical reaction is like respiration in reverse. Write a word equation for this reaction.
3. Yeast is a microorganism used in brewing and baking. It changes sugar into ethanol (alcohol) and carbon dioxide as it respires. This process is called fermentation.
   - **a** Write a word equation for fermentation.
   - **b** Which is the useful product in brewing?
   - **c** Which is the useful product in baking? Explain your answer.
4. Plants also need phosphorus to grow well. Ammonium phosphate is made by reacting ammonium hydroxide with phosphoric acid. Write a word equation for this reaction.

Did you know?

Plants make food using the chemical reaction called photosynthesis (see page 22). Plants and animals get energy from food using the chemical reaction called respiration. Respiration is the opposite reaction to photosynthesis.

For your notes:

- Most materials around us are made by chemical reactions.
- Some chemical reactions are useful and some are not.

H2 Energy from reactions

Learn about:

- Energy from chemical reactions

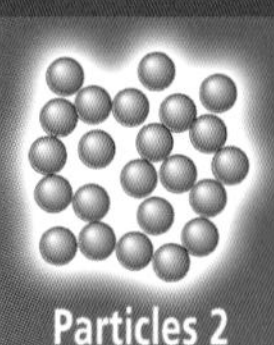

Getting energy

Thermal energy is often released during chemical reactions. Combustion and displacement reactions are two examples. In the displacement reaction, when you add zinc to a test tube of copper sulphate solution, you will feel it warm up. If you add magnesium instead, more energy is released. The further apart the two metals are in the reactivity series, the more thermal energy is released.

When you place two different metals in a dilute acid, electrical energy is produced. An electrical current flows between the two metals. Again, the further apart the two metals are in the reactivity series, the more thermal energy is released. This is a simple battery, or **voltaic cell**, shown in this diagram.

01:50
volts
acid
copper
zinc

Fuels and combustion

Combustion reactions release thermal and light energy, which is why they are so useful to us. Fuels contain stored chemical energy. This is released when they burn. Combustion is the chemical reaction of a fuel with oxygen in the air. For fuels, the energy released is more important than the new chemical products.

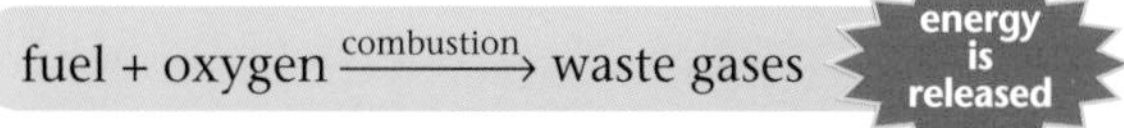
fuel + oxygen $\xrightarrow{\text{combustion}}$ waste gases — energy is released

Do you remember?

Fossil fuels formed from the remains of plants or animals. The plants originally got their stored energy from the Sun.

Gas is another form of fossil sunshine.

Burning fossil fuels

Some people burn wood to heat their homes. But most people in Britain rely on fossil fuels: coal, oil and gas. Even if your home is 'all electric', coal is burned in many power stations to generate the electricity. Coal is an impure form of the element carbon, which produces carbon dioxide when it burns.

carbon + oxygen → carbon dioxide — energy is released

Oil and gas are **hydrocarbons**, which are compounds containing carbon and hydrogen. Natural gas contains methane.

methane + oxygen → carbon dioxide + water — energy is released

a **What are the products when all hydrocarbons burn?**

Some less useful products

Burning fuels in cars makes some harmful products, which lead to acid rain or smog. They also make carbon dioxide. This gas is not usually thought harmful, but we make so much that it might cause global warming (see page 70).

Cars also make deadly poisonous carbon monoxide. Carbon monoxide is made when there is not enough oxygen to react with the fuel. It is also made if you have a faulty gas heater at home. If you breathe in carbon monoxide, it stops your blood carrying oxygen and could kill you.

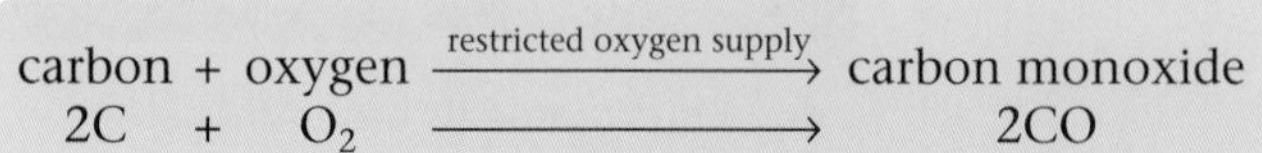
carbon + oxygen $\xrightarrow{\text{restricted oxygen supply}}$ carbon monoxide

$2C + O_2 \longrightarrow 2CO$

b **The flame in a gas heater should look like a little roaring Bunsen flame. If it gets yellow and sooty, like in the photo on the right, not enough oxygen is reacting. Why is this dangerous?**

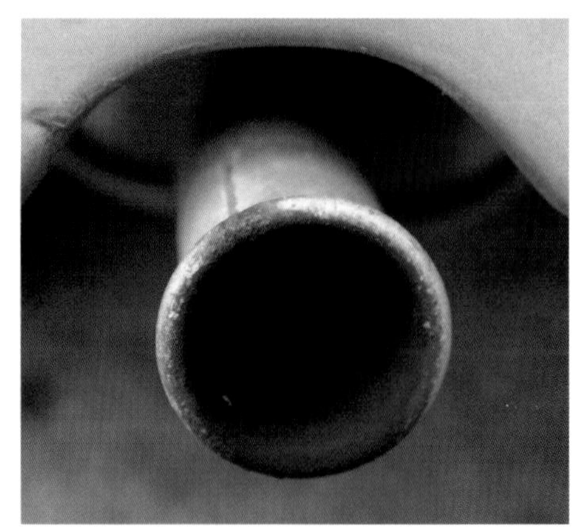

This black soot is caused by carbon build-up in the car's exhaust.

Alternative fuels

There's one big problem with fossil fuels. They are going to run out, even if we try to conserve them to make them last longer. We can make cars with smaller engines, which burn less petrol and travel further per litre of petrol. But we still need to find alternative fuels.

In Brazil, where there is little fossil fuel but lots of sugar cane, cars run on ethanol instead of petrol. The ethanol is made from sugar.

c Write a word equation for burning ethanol.

As sugar cane can be grown year after year, ethanol will not run out. It is a renewable biomass fuel that relies on living organisms.

The space shuttle burns hydrogen in its main engines.

Hydrogen – the fuel for the future

Hydrogen may be a fuel for the future. It is already used in space programmes to power rockets.

hydrogen + oxygen → water

d Why is hydrogen a cleaner fuel than ethanol?

There are problems with using hydrogen as a fuel. For example, hydrogen is produced by splitting up water using electricity. If we use 'mains' electricity produced from fossils fuels, the pollution at the power station cancels out the pollution saved by using hydrogen-powered cars. All this does is move the pollution from the roads to the power stations.

One solution is to use solar cells, which make electricity directly from sunlight. Using solar cells to produce electricity for making hydrogen really would be a 'pollution-free' fuel. Unfortunately, large solar cells are very expensive and it isn't always sunny.

Do you remember?

Hydrogen is a gas, which takes up a lot of space, so it has to be compressed to a liquid in a very strong container. Also, hydrogen can explode when it is mixed with oxygen.

Questions

1 When wood burns in oxygen it forms carbon dioxide and water. Which two elements must it contain?

2 Steam trains were coal fired. Modern trains often have diesel engines. The trains of the future could run on hydrogen. Explain how improving railway technology would be better for the environment.

3 Explain why hydrogen might not be quite so good for the environment if it is made using electricity from an oil-fired power station. How do solar cells overcome this problem?

4 An exploration satellite may have detected ice deposits on the Moon. Explain how this could be used to help colonise the Moon. (*Hint*: the cost of solar cells would not be too important on the Moon, when compared with the cost of getting there.)

For your notes:

- **Voltaic cells** and many chemical reactions, such as displacement and combustion reactions, release energy.
- Coal produces carbon dioxide when it burns.
- **Hydrocarbons** produce carbon dioxide and water when they burn.
- Hydrogen is a 'pollution-free' fuel as it produces only water when it burns.

H3 Reactions in balance

Learn about:
- Conserving mass
- Balanced equations

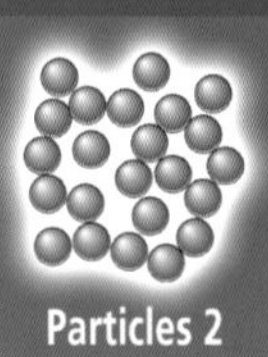

Making a compound

Kerry and Nathan made iron sulphide by heating iron filings and sulphur in a test tube. They talked about how much iron sulphide was made.

a **Who do you agree with, Kerry or Nathan? Explain your answer.**

The reactants are iron atoms and sulphur atoms. The product iron sulphide contains both iron and sulphur atoms combined. Each iron atom is next to sulphur atoms, and each sulphur atom is next to iron atoms.

The total number of atoms in the reactants and the product is the same. They are just rearranged into a compound. We can write a word equation and a balanced equation for the reaction.

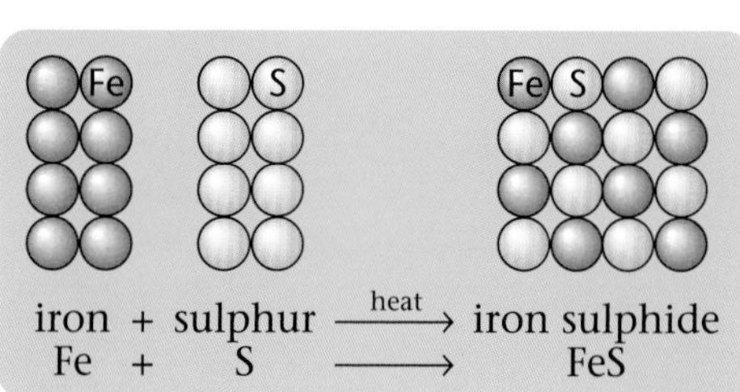

Do you remember?

The mass of a solution is the same as the mass of the solute and solvent together.

The mass stays the same

In all chemical reactions, the total number of atoms stays the same. So the total mass stays the same – we say that mass is **conserved**.

These two photos show hydrochloric acid neutralising sodium hydroxide. The products are sodium chloride (common salt) and water.

b **Look at the photos. Explain why both balances show the same reading.**

Here are the word equation and chemical formulae of the compounds involved.

hydrochloric acid	+	sodium hydroxide	→	sodium chloride	+	water
HCl	+	NaOH	→	NaCl	+	H_2O

c **Add up each kind of atom on each side of the equation to see if it balances.**

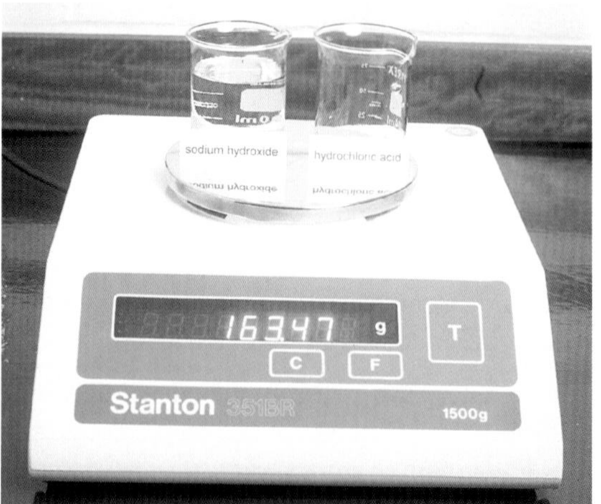

Before and after mixing.

How to balance equations

A balanced equation has the same number of each type of atom on both sides. Some equations are balanced when you first write them:

$HCl + NaOH \rightarrow NaCl + H_2O$

Others are trickier. But you can balance equations if you follow these simple rules ...

1 Write a word equation for the reaction. In the example on the right, hydrogen is burning in oxygen.

2 Write the formula of each reactant and product. Both hydrogen and oxygen are molecules with two atoms each.

3 Count the atoms on each side. Do you have the same number of each type on both sides? In this case, no!

4 There are two oxygen atoms on the left, but only one on the right. You need an extra water molecule on the right. You show this in the equation by putting a 2 in front of the formula for water. This means '2 molecules of'.

5 Does it balance? Again, no! This time you have 4 hydrogen atoms to the right and only 2 to the left. You will need to start with 2 molecules of hydrogen.

6 Now it balances.

Example

hydrogen + oxygen → water

$H_2 + O_2 \rightarrow H_2O$

$H_2 + O_2 \rightarrow 2H_2O$

$2H_2 + O_2 \rightarrow 2H_2O$

Questions

1 When 5.6 g of iron reacted with exactly the right amount of sulphur, it produced 8.8 g of iron sulphide. What mass of sulphur was used?

2 When you heat magnesium ribbon in air it burns with oxygen (O_2) to give white magnesium oxide (MgO). If you burn 2.4 g of magnesium, you get 4 g of magnesium oxide.

- **a** Write a word equation for the reaction.
- **b** Why does the mass seem to go up?
- **c** What mass of oxygen must have combined with the magnesium?
- **d** Write a balanced equation for the reaction.
- **e** If you burn 1.2 g of magnesium, how much magnesium oxide would you get? Explain your answer.

3 If you heat calcium carbonate ($CaCO_3$) it breaks down into calcium oxide (CaO) and carbon dioxide (CO_2). 100 g of calcium carbonate gives just 56 g of calcium oxide.

- **a** Write a word equation and a balanced equation for the reaction.
- **b** Does the equation balance when you first write it?
- **c** Why does the mass go down in this reaction?

4 Write a balanced equation for each of these reactions.

- **a** Sodium (Na) reacts with chlorine gas (Cl_2) to produce sodium chloride ($NaCl$).
- **b** Sodium (Na) reacts with water (H_2O) to produce sodium hydroxide ($NaOH$) and hydrogen (H_2).

Do you remember?

When balancing equations:

- you *cannot* change the *small* numbers *within* the formulae; that would change the substance
- you *can* change the *large* numbers in *front* of the formulae, to make the equation balance.

For your notes:

- In a chemical reaction, mass is **conserved** because the total number of atoms stays the same.
- In a chemical reaction, the atoms are rearranged to make new compounds.

H4 The story of burning

Think about:

- What happens when things burn …
- … and how we found out

Phloginston rules!

Today, we know that when something burns, it reacts with the oxygen in the air. How was this discovered?

When wood burns you are left with ash – a lot less ash than the wood you started with. About 300 years ago, a scientist called George Stahl came up with the idea of **phlogiston** to explain burning.

You can show this hypothesis by writing:

fuel → ash + phlogiston (in the flames)

a **How does the idea that something is lost on burning compare with our modern understanding of what happens?**

The idea of phlogiston was very popular for a long time. Scientists used it to explain burning.

A new discovery

In 1774, Joseph Priestley did an experiment heating a substance called red calx. He knew that this was used to make mercury. He heated it using a lens to focus the rays of the Sun. Mercury was made, as he expected. But he was surprised that a gas was also made.

red calx → mercury + new gas

Priestly was even more surprised when he found that this new gas made a candle burn more brightly.

An alternative idea

A year or so later, Antoine Lavoisier decided to reverse Priestley's experiment. He burned mercury to see what happened. Lavoisier did his burning experiment very carefully. He trapped the mercury and the air in a special flask, with its end dipped into a bowl of liquid mercury.

before

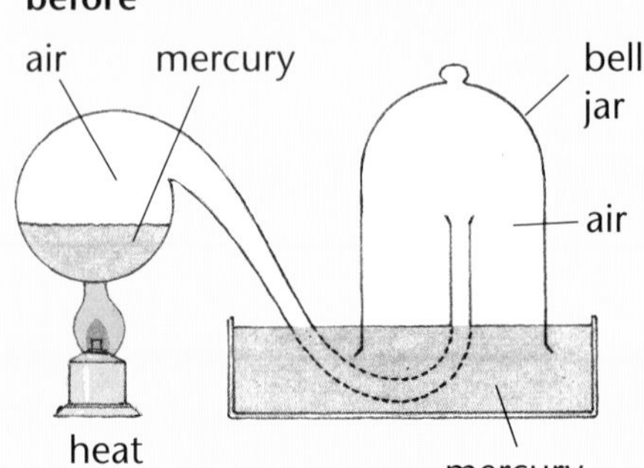

Lavoisier heated the mercury in the flask. A red substance that looked like red calx was made on its surface. He also noticed that some of the mercury in the bowl was sucked into the flask. This meant the amount of air in the flask had decreased! Something had *left* the air and joined with the mercury.

mercury + new gas → red calx

after

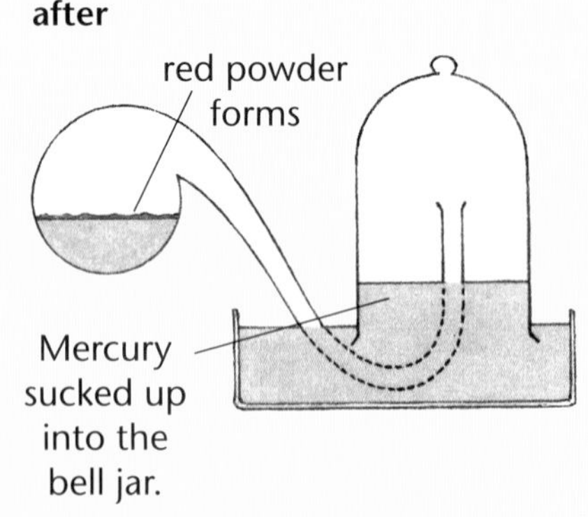

Lavoisier had something *joining* during burning. Stahl and Priestley had something (phlogiston, or gas) *leaving* during burning.

b **Which idea matches what we know today?**

Lavoisier realised that the gas taken out of the air in his experiment must be the same as Priestley's 'new gas'. He named this gas oxygen. Then he showed that oxygen was involved in burning.

mercury + oxygen → red calx

A modern explanation

When mercury is heated in air, mercury atoms combine with oxygen atoms in the air to form a compound called mercury oxide. Mercury oxide is the scientific name for the substance once called red calx. You can write a word equation for Lavoisier's experiment:

mercury + oxygen → mercury oxide

c **Write a word equation for Priestley's experiment heating mercury oxide.**

Convincing everyone

The idea of phlogiston was accepted among scientists. It seemed to fit some cases. For example, when wood is burned, the ash left weighs much less than the wood. It fits with the idea of something being lost.

d **How can you explain this 'mass loss' when wood is burned, using what we know today?**

To convince other scientists, Lavoisier had to show that his idea was better at predicting what would happen when other substances were burned. Lavoisier did many experiments. In one, he carefully weighed some tin, burned it to leave a white substance, and then reweighed this new substance.

e **Would scientists who followed the phlogiston idea have expected the mass of the new substance to be greater than or less than that of the original tin?**

Lavoisier's hypothesis predicted that the metal would combine with oxygen from the air when it reacted, and so the mass would go up. His experiments showed that the mass did indeed go up, not down. This supported his idea and proved that the phlogiston idea was wrong. Eventually, all scientists agreed with Lavoisier and his 'new idea' about oxygen became the accepted theory.

Antoine Lavoisier.

Questions

1 Tony wanted to try a modern version of Lavoisier's experiment. He put a crucible, on top of a heatproof mat, on an electronic balance connected to his computer. He used datalogging to take a mass reading every second. He put a coil of magnesium ribbon into the crucible and lit it. This graph shows his results.

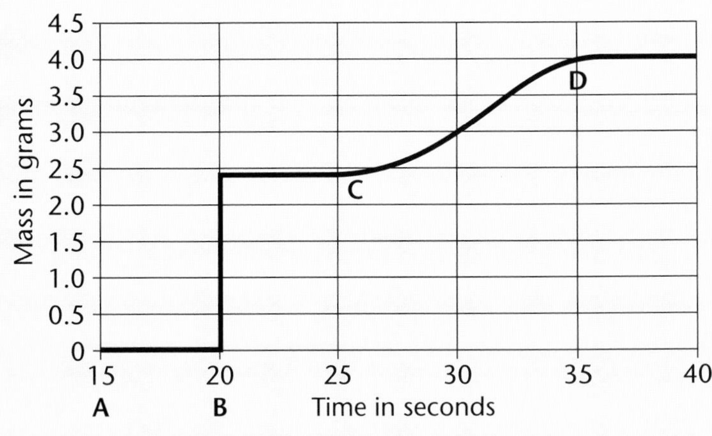

a What must be happening at points B,C and D shown on the graph?

b Describe how this graph supports Lavoisier's idea.

c Tony repeated the experiment with 2.4 g of wood instead of the magnesium. Sketch the graph that you would expect to get from this.

d This second graph could be used to support the phlogiston theory. How could you adapt this experiment to show that the products of combustion are actually being lost as gases?

e Tony read about another experiment on the Internet. In a research laboratory, 2.4 g of charcoal (carbon) were burnt in oxygen and the gas produced was sucked through another tube containing a chemical that absorbed carbon dioxide. At the end, this second tube was found to have increased in mass by 8.8 g. Explain how this experiment confirms Lavoisier's ideas and disproves the phlogiston idea.

I1 Make it work

Learn about:

- Energy transfers involving electricity
- Conservation of energy

On the move

Ellen's MP3 player works because of electricity. It takes in electrical energy from the batteries and gives out sound energy. The energy transfer diagram on the right shows this.

electrical energy → MP3 player → sound energy

The microwave oven works because it is plugged in. It gets the electrical energy it needs to work from the mains. This energy transfer diagram shows this. Electricity is a clean and convenient source of energy.

electrical energy → microwave oven → thermal (heat) energy; kinetic (movement) energy; light energy; sound energy

The energy given out by the MP3 player and the microwave oven came from the electricity. Energy always comes from somewhere and goes somewhere. It is not created and it is not destroyed: it is just moved around. We say that energy is **conserved**.

In and out of storage

This photo is from Ellen's skiing holiday.

Look at the energy transfer diagram for the ski lift. It takes in electrical energy and lifts the skiers up. The skiers end up with a store of gravitational energy. They then use this stored energy to ski down the slope.

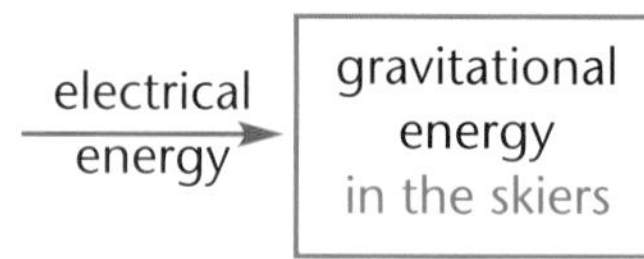

Energy can be stored, as well as being transferred, but it is still conserved.

a **The ski lift in this photo has an electric motor, which gets its energy from the mains. Many ski lifts have a diesel motor, which burns fuel to release energy.**
(i) Suggest an advantage of the electric motor.
(ii) Suggest an advantage of the diesel motor.

Did you know?

Gravitational energy is also called **gravitational potential energy**.

Ellen wants to buy a torch for her mother to keep in the car for emergencies.

- One torch has batteries. It has a store of chemical energy in the batteries.
- The other torch is a wind-up torch like the one in the photo on the left. You wind the handle and the torch stores energy as strain energy. When you switch the torch on, the electricity carries the energy between the store and the lamp, making it work.

The wind-up torch costs £39.99 and the battery torch costs £9.99.

b **Draw an energy transfer diagram for each torch when it is switched on. (*Hint*: use the ski lift diagram to help you.)**

c **Which torch would you buy? Give one reason why you chose one torch and one reason why you did not choose the other.**

Keeping track

You can look at the energy before and after an energy transfer and show it is conserved. Think about a hairdryer. 750J of electrical energy goes into the hairdryer. 735J of thermal energy, 5J of sound energy and 10J of kinetic energy come out of the hairdryer.

energy in = 750J
energy out = 735J + 5J + 10J = 750J

Of the 750J, 745J ends up somewhere useful. To dry hair you need heated, moving air but you do not need sound. About 99.3% of the energy is useful; the other 0.7% is wasted.

d An electric lamp takes in electrical energy and gives out light energy and thermal energy. 100J is taken in every second. 30J is given out as light energy every second.
(i) How much energy is given out as thermal energy?
(ii) What percentage of the energy ends up where it is wanted, as light energy?

Do you remember?

Energy is measured in joules, J. There are 1000 joules in one kilojoule, kJ.

Spread about

Think about the electric torch. The light energy leaves the torch. Where does it end up? Some of the energy ends up in your eye, making you see. Some of the energy is carried far away, reflected from surface to surface. Some of the energy is absorbed when the light hits materials and is not reflected. The particles in the material vibrate slightly more when the light is absorbed, raising the temperature of the material by a tiny amount. We say that the energy is spread about or **dissipated**.

£10 is a very useful amount of money. One person with £10 can buy a lot of different things. £10 spread between 1000 people is a lot less useful. Each person has 1p. You can't buy much with 1p. Once the money has been dissipated it is a lot less useful.

Energy is like money in this way. As it is spread about it becomes less useful.

Questions

1 Draw an energy transfer diagram for:
- **a** a hairdryer
- **b** a wind-up clock.

2 A TV takes in 2000J every second. It gives out 200J as light energy and 50J as sound energy every second. It also heats up.
- **a** How much energy per second does the TV give out as thermal energy?
- **b** Where does this energy end up?
- **c** How much energy per second ends up where you want it, as light energy and sound energy?
- **d** Explain how you used the idea of **energy conservation** when answering this question.
- **e** What example of **energy dissipation** is given in this question?

3 People lived for thousands of years without electricity. How would your life be different without electricity?

For your notes:

- Energy is transferred as electrical energy, sound energy, light energy, thermal energy or kinetic energy.
- Energy is stored as chemical energy, gravitational energy or strain energy.
- Electricity is a clean and convenient source of energy.
- Energy is **conserved**.
- Energy is spread about or **dissipated**.

I2 Energy in and out

Learn about:

- Voltage

Circuits

We can use Ellen on her skiing holiday as a model of an electric circuit.

The flow of the skiers in the diagram above shows the current. Energy is put into the circuit at the ski lift, where the skiers are lifted up. Energy comes out of the circuit at the ski run, as the skiers zoom down the slope. The ski lift is like the cell, putting energy into the circuit. The downhill ski run is like the lamp, where the energy comes out of the circuit.

Using this model we can show that energy is conserved. The skiers gain a store of energy as they are lifted up, and transfer it away as they ski down. In the same way, energy is put into the circuit at the cell, and transferred away from the circuit as the current flows through the other components.

We can also apply the idea of energy dissipation to this model. What happens to the kinetic energy that the skiers had when they zoomed down the slope? It was dissipated to the surroundings, heating up the air and the snow. In the same way, energy is dissipated from a circuit with a lamp. A small fraction of it leaves the circuit as light energy. The rest is dissipated as thermal energy.

We can also introduce the idea of electrical resistance into this model. The snow on the ski can be rougher or smoother. If it is rougher, there will be more friction and the skiers will move more slowly down the slope. The rougher snow is like a material with a higher resistance. The flow of skiers is slower, just like the current is lower in a material with a greater resistance.

Do you remember?

Some materials have a high electrical resistance. They don't conduct electricity very well. Current flows less easily through high-resistance materials.

Voltage

The cell in this photo has a voltage of 1.5 volts, 1.5 V for short. We measure voltage with a **voltmeter**. The circuit symbol for a voltmeter is (V).

Mel and Jeff are measuring the voltage across four parts of a circuit. A drawing of their experiment is shown opposite, with the circuit diagram beside it. The voltmeter is not in the circuit. It is in a separate loop. You connect the voltmeter across the battery or across the lamp. The black wires show the circuit. The wires going to the voltmeters are pink.

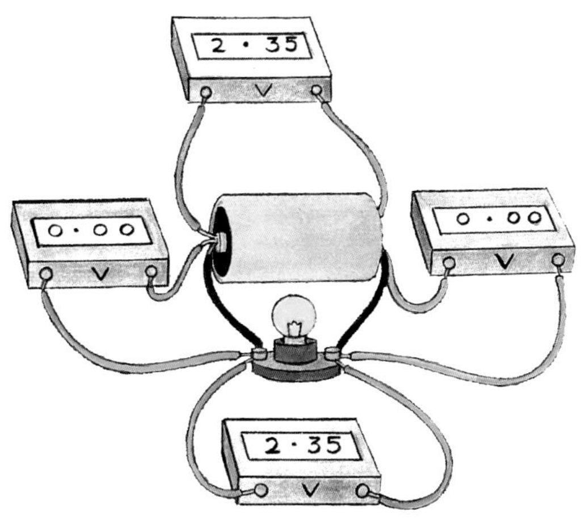

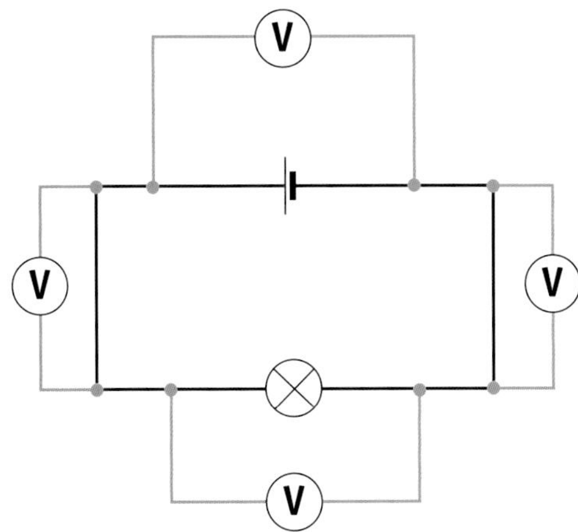

Voltage tells us where there is a change in energy in the circuit. There is a voltage across the cell: this is where energy is put in. There is a voltage across the lamp: this is where the energy leaves.

a **Why is there no voltage across the wires?**

Modelling voltage

Where does voltage fit into our model of the circuit?

Look back at the model. The skiers at the bottom have less energy because they are lower down. The skiers at the top have more energy because they have been lifted up. The voltage is the change in height. Another term for voltage is **potential difference**. This refers to the difference in energy.

There is a big height change over the ski lift part of the circuit. This shows the voltage across the cell. There is a big height change when the skiers go down the ski run. This shows the voltage across the lamp. But there is no height change when the skiers walk along at the top and the bottom. This shows that there is no voltage across the wires.

b **What is another scientific term for voltage?**

Questions

1. **a** Where in a circuit do you measure a voltage?
   b What do you use to measure voltage?
   c What are the units of voltage?
2. Think about the ski resort model of an electric circuit.
   a What is represented by:
   (i) the ski lift? **(ii)** the downhill ski run?
   (iii) the flow of skiers? **(iv)** the height change?
   (v) the quality of the snow?
   b Energy is conserved in an electric circuit. How does the ski resort model show this?
   c Energy is dissipated in an electric circuit. How does the ski resort model show this?
3. Using the ski resort model, describe a model of a circuit with two cells and three lamps in series.
4. The energy transferred in a circuit depends on both the voltage and the current. Use the ski resort model to explain this.

Energy in circuits

Think about the model again. Energy is put into the circuit when the skiers are lifted up. The bigger the change in height, the more energy is put into the circuit. The more skiers that are lifted up, the more energy is put into the circuit. The amount of energy put in depends on two things, change in height and the number of skiers using the ski lift. The change in height represents the voltage. The flow of skiers through the ski lift represents the current. The total energy put into the circuit depends on both voltage and current.

For your notes:

- Voltage is measured across parts of a circuit, using a **voltmeter**.
- Voltage is also called **potential difference**.
- There is a voltage across any part of the circuit where energy is coming in or going out.
- The energy transferred in a circuit depends on both voltage and current.

13 Using electricity

Learn about:

- The dangers and benefits of electricity
- Saving energy

High voltage

Look at the photo of lightning on the right. The voltage between the cloud and the ground is 300 000 V. That means a lot of energy is being transferred between the cloud and the ground.

Fast-moving electric trains have lots of kinetic energy. The overhead cables in the photo below have a voltage of 25 000 V across them. The mains cables coming into our homes have a voltage of 230 V across them. This voltage is enough to give us the electrical energy we need. The higher the voltage, the more energy we can transfer from the circuit.

a **Why can't you run electric trains on a 230 V mains supply?**

Did you know?

Electricity kills about 30 people a year in UK homes and injures about 2000 others. Treat it with respect: never handle electrical devices with wet hands, keep them away from water, and always use a qualified electrician to do repairs.

Paying for energy

The electricity meter in your home measures the amount of energy you transfer from the mains. You pay the electricity supplier for the energy.

Different devices transfer different amounts of energy from the mains. An electric cooker or a washing machine needs much more energy than an electric lamp. Even different electric lamps transfer different amounts of energy from the mains. A 100 watt bulb transfers 100 joules of energy from the mains every second. A 60 watt bulb only transfers 60 joules per second. We say the 100 watt bulb has a **power rating** of 100 W and the 60 watt bulb has a power rating of 60 W.

This table shows the power ratings of a range of electrical devices we use at home.

Electrical device	Power rating in watts (W)
100 watt light bulb	100
60 watt light bulb	60
electric cooker	132 000
dishwasher (hot cycle)	1450
dishwasher (cool cycle)	700
microwave oven	850
washing machine	1200
tumble dryer	5800

b **Explain how heating food in a microwave rather than an electric oven saves money.**

We need to save energy for other reasons as well as saving money. We use non-renewable energy sources, such as the burning of fossil fuels, to make most of our electricity. We need to save energy to make these last as long as possible. Saving energy would also mean we produce less polluting gases, causing less global warming and acid rain.

Do you remember?

Fossil fuels are non-renewable and contribute to global pollution.

Same voltage, different energy

Everything we plug in connects to a circuit with a voltage of 230 V. It is true of a lamp, giving out 60 joules per second, and a kettle, giving out 3000 joules per second. How can the same voltage deliver different amounts of energy?

The answer is that the energy transferred depends on the current as well as the voltage. The circuit through a table lamp will have a low current, less than 3 A. The electric cooker may have a current of up to 30 A.

Energy efficiency

Look at these labels from two washing machines. One has a high 'A' grade for **energy efficiency**. The other has only a 'B' grade. Devices with a high energy efficiency waste less energy.

When you transfer energy away from the mains, not all of it ends up where you want it. Think about a washing machine. You want the energy to be turning the drum of the washing machine, pumping the water in and out. You don't want the energy to be warming up the washing machine and the air around it, making the washing machine vibrate and producing sound. The 'A' grade washing machine wastes less energy. More of the energy ends up where you want it to go.

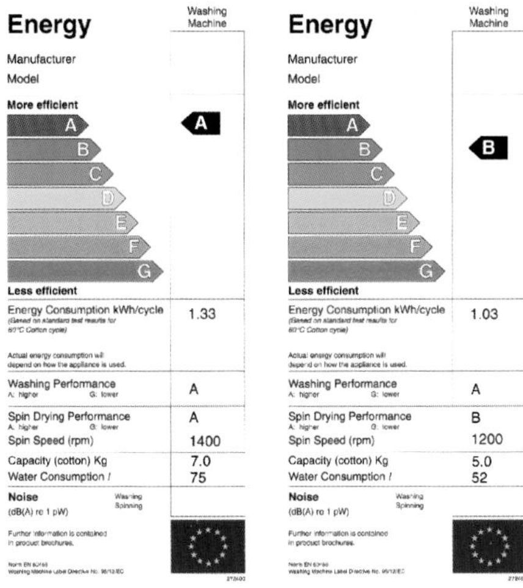

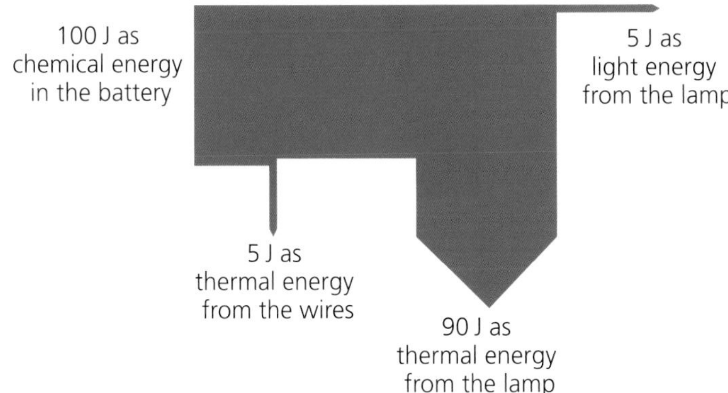

We cannot stop energy being dissipated, or spread about. But we can try to make sure that more of the energy ends up where we want it. Devices with a high energy efficiency dissipate less energy to the surroundings.

We can show how much energy ends up were we want it using a **Sankey diagram**. Look at this Sankey diagram on the left for a battery-powered torch. The width of the arrow represents the chemical energy in the battery. The horizontal line represents the amount of energy that ends up where you want it, in this case as light energy. Only 5 J out of every 100 J ends up as light. That's just 5%. The other 95 J, or 95%, is dissipated to the surroundings as thermal energy.

Questions

1 Use the information in the table of power ratings to explain how you save energy if you:

- **a** switch off lights
- **b** change 100 W bulbs to 60 W bulbs
- **c** hang out your washing rather than use a tumble dryer
- **d** run your dishwasher on the cool cycle.

2 Helen is trying to persuade her mother to buy a more energy-efficient washing machine. The more energy-efficient washing machine costs £65 more. Explain how the more expensive washing machine could save Helen's mother money.

3 Joe has electric storage heaters to heat his home. Explain how insulating his house can:

- **a** save Joe money
- **b** make fossil fuels last longer
- **c** help save the planet!

For your notes:

- The higher the voltage, the more energy can be transferred.
- Different devices transfer different amounts of energy from a circuit.
- We pay for electrical energy, so using less electricity saves money and energy.
- Devices with high **energy efficiencies** dissipate less energy to the surroundings.

I4 Power stations

Learn about:

- Transferring energy to electricity

Energy in

We use electricity to transfer energy into our homes. How does the energy get into the electricity to start with?

Look at this bike. It has a dynamo. The dynamo transfers energy from the moving wheel to electricity, which then lights the bike lights. A dynamo takes in kinetic energy and produces electrical energy. The electricity we use at home is made in power stations that have huge dynamos called **generators**, like the one in the photo below.

It took the great experimental scientist Michael Faraday 10 years to come up with an apparatus that used movement to produce electricity. He succeeded in 1832. All modern dynamos and generators are based on Faraday's experiment. You spin a magnet to make a changing magnetic field. If you put a wire in the changing magnetic field, a current flows along the wire.

Turning the generator

On the bike, the wheel spins the dynamo to make electricity. The generator also has to spin to produce electricity. Generators are spun by a **turbine**. Most generators are spun by a steam turbine. Water is heated until it boils. The steam is then used to turn the turbine.

The diagram below shows how a steam turbine works. Read the labels carefully. The numbers refer to the energy transfer diagram shown below. The two diagrams describe the energy transfers that happen in the power station.

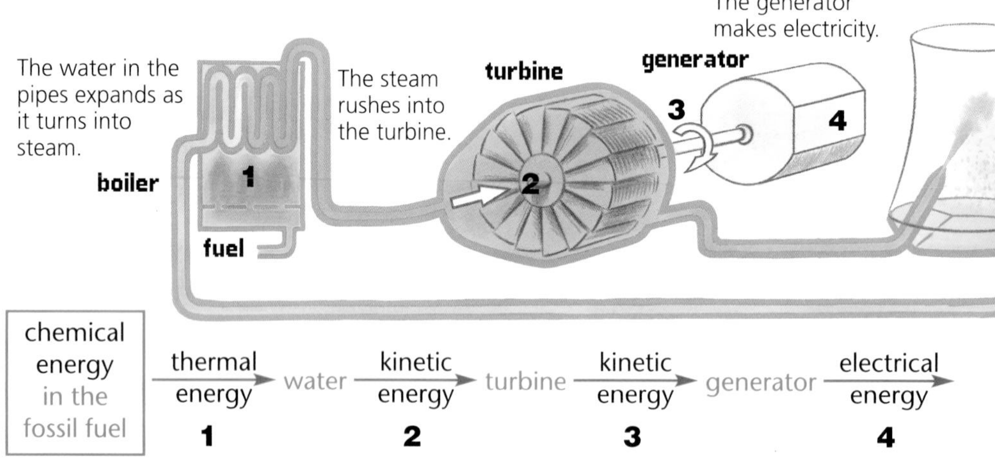

chemical energy in the fossil fuel → thermal energy (1) → water → kinetic energy (2) → turbine → kinetic energy (3) → generator → electrical energy (4)

The water can be heated in many different ways. Many power stations in Britain burn natural gas to heat the water. Others burn coal or oil. In hot, sunny countries the water can be heated to boiling by the Sun.

a **Draw your own energy transfer diagram for a power station that uses solar power.**

Power stations have to supply people's demand for electricity at the very moment they need it. During big World Cup football matches the demand for electricity goes up at half-time, when people switch their kettles on across the whole country. That means more generators have to be spinning. Electrical energy cannot be stored. It is energy on the move.

Do you remember?

Many energy resources can be used to generate electricity. These include renewable energy resources such as biomass, wind, waves, solar and falling water, and non-renewable resources such as fossil fuels.

How efficient?

Look at the Sankey diagram on the right showing the energy transfers through a fossil fuel power station. Only 33% of the energy in the fuel ends up as electrical energy. The rest is dissipated as thermal energy to the surroundings.

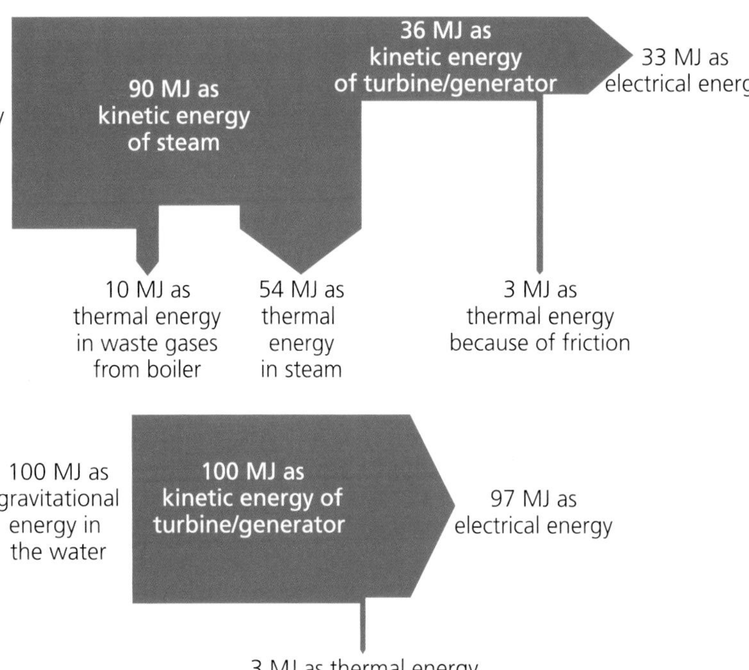

b **Some power stations use the hot water from the cooling towers to heat nearby homes. Explain how this helps conserve fossil fuels.**

Compare the Sankey diagram for the fossil fuel power station with the one for a hydroelectric power station, shown on the right.

c **How much energy is dissipated to the surroundings by the hydroelectric power station?**

Pros and cons

The different energy resources we use to generate electricity all have different advantages and disadvantages. Some of these are summarised in this table.

Energy resource	Fossil fuels	Biomass	Solar	Wind	Waves	Falling water
renewable?	no	yes	yes	yes	yes	yes
causes acid rain?	yes	no	no	no	no	no
causes global warming?	yes	yes	no	no	no	no
energy efficient?	no	no	no	yes	yes	yes
expensive to build?	less	less	more	more	very	more
expensive to operate?	yes	yes	no	no	no	no
destroys habitats when built?	yes	yes	yes	yes	probably	yes
noisy?	no	no	no	very	no	no
ugly?	yes	yes	yes	yes	yes	yes
restricted sites to build generators?	no	no	yes	yes	yes	yes

Questions

1 Write definitions for each of these scientific words. Then check your definition against the one given in the glossary.

conserved **dissipated** **energy efficiency** **generator** **turbine**

2 Imagine you are each of these people in turn. Use your knowledge about energy resources and the information in the table to decide which power station you would want built. Give the three main reasons for each choice.

- **a** The government minister in charge of the long-term strategy for electricity generation.
- **b** A forester whose livelihood is threatened by acid rain.
- **c** A scientist who believes that global warming will radically change the world's climate within 20 years.
- **d** Someone who wants to pay less for electricity in the short term.
- **e** A person living next to the proposed power station.

For your notes:

- **Generators** transfer energy to electricity.
- **Turbines** transfer kinetic energy to generators.
- Turbines can be turned by steam, wind, falling water and waves.
- The steam is often made by burning fossil fuels. The waste gases from this cause global warming and acid rain.
- Fossil fuels are non-renewable. Solar power, wind, waves and falling water are renewable.

J1 A massive problem

Learn about:

- Mass, distance and gravitational attraction

Attracting bodies

It is difficult to believe but anything that has mass is attracted to anything else with mass. You have mass, so you are attracted to your pen, this book and the person next to you, because they all have mass. The more mass objects have, the more they are attracted. You are most attracted to the Earth, which is the most massive object around.

a What would Joe weigh if you could get him to a place where there was no gravitational attraction?

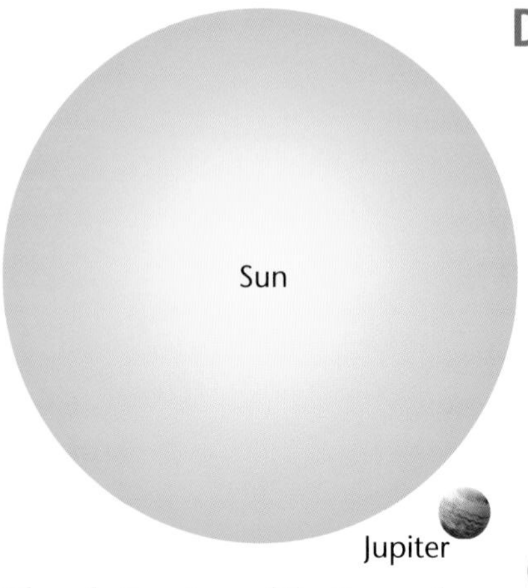

The relative sizes of the Sun and Jupiter. On this scale, the Earth would have a diameter of 1 mm.

Different places

Different planets have different gravitational attraction. The gravitational attraction of Venus is 9 N/kg, a bit smaller than the Earth's 10 N/kg. This is because Venus has smaller mass than Earth. Jupiter is the most massive planet in the Solar System. It has a gravitational attraction of 26 N/kg.

b Joe has a mass of 50 kg. How much would he weigh on: (i) Jupiter? (ii) Venus?

Do you remember?

Each kilogram of mass is attracted to the centre of the Earth by a force of 10 newtons. On the Moon, gravitational attraction on a mass of 1 kg is 1.7 N.

There is a gravitational attraction between any object and the Earth. We call that force weight. The Earth has a gravitational attraction of 10 N/kg, or 10 N on every kilogram.

Do you remember?

The Solar System has the Sun at its centre. Nine planets orbit the Sun.

The Sun has a huge gravitational attraction because the Sun has a huge mass of 2 000 000 000 000 000 000 000 000 000 000 kg. A mass of 1 kg would weigh 279 N on the surface of the Sun. This huge gravitational attraction keeps all the planets in their orbits.

c How much would Joe weigh if he could stand on the surface of the Sun?

Escaping the Earth

It's a lot harder to get more massive objects away from the Earth's surface than small objects. Think about what happens when you jump. You push against the Earth and you move upwards, but the Earth is always pulling you back down. To keep going upwards, you would have to keep pushing.

That's what rockets do. They produce a steady **thrust** or pushing force. To start moving upwards, the thrust has to be greater than the weight of the rocket.

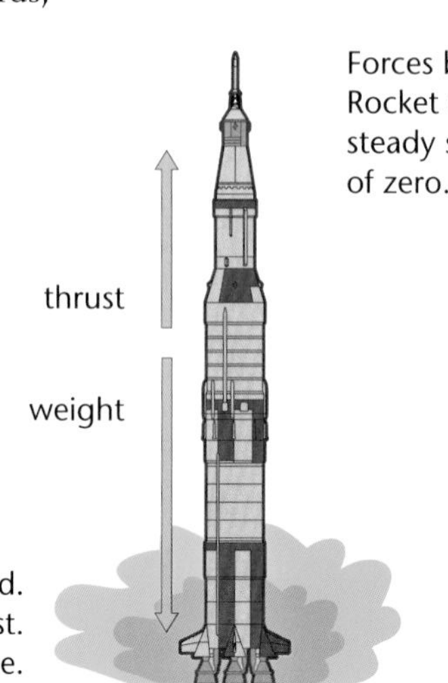

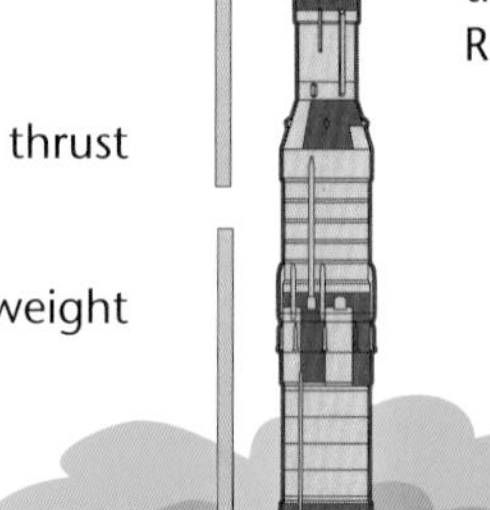

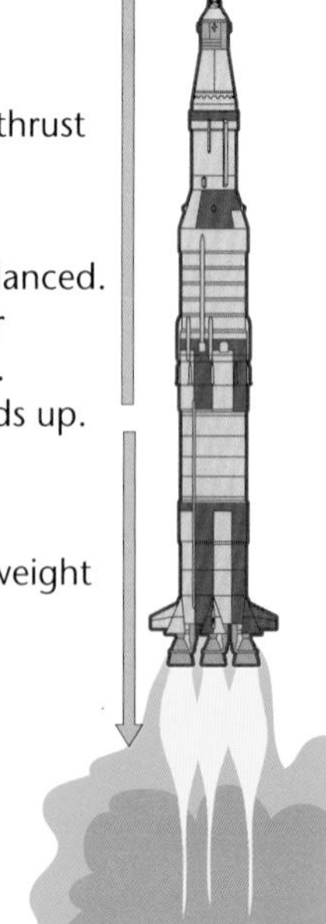

The problem is that the engines to produce all this thrust are huge. Look at this photo. It shows a Saturn V rocket. It has a huge mass and it also needs to carry a huge mass of fuel to keep the engines working. But all this mass means the rocket has a huge weight, which is hard to overcome.

d **As the rocket travels upwards, it burns fuel and expels the waste gases.**
(i) How will this change the weight of the rocket?
(ii) How will this change the amount of thrust needed to move the rocket?

Thinking about gravity

It took the amazing imagination of Isaac Newton to understand how gravitational attraction worked. He took data that other people had collected and analysed it in many different ways. He came up with a law of gravity that has two parts.

The first part is that the gravitational attraction between two objects depends on the mass of each object, as described above. The second part is about distance. The further the objects are apart, the weaker the gravitational attraction between them.

Using Newton's law

We can use Newton's law to show what would happen if a rocket with a mass of 1000 kg escaped the Earth and made a trip to the Moon.

e **Explain why position 2, where the two forces balance, is closer to the Moon than to the Earth.**

f **Explain why the force between the rocket and the Earth is 10 000 N on the Earth and only 3.5 N when it is 354 000 km from Earth.**

Questions

1 The Saturn V rocket, fully fuelled, had a mass of 3 038 500 kg.

a What was the weight of the rocket?

The lift-off thrust was 33 737 900 N.

b How much of this thrust was needed to balance the weight of the rocket?

c How much of this thrust was pushing the rocket upwards?

2 Describe the relationship between the gravitational attraction between two objects and the distance between them.

3 After launch, a rocket moves upwards, the fuel is used and the waste gases are expelled.

a How is the mass of the rocket changing?

b How is the distance from the centre of the Earth changing?

c What effect will changing the values of mass and distance have on the weight of the rocket?

For your notes:

- Gravitational attraction depends on the mass of the two objects being attracted. It is different on other planets and on the Moon. The Earth's gravitational attraction is 10 N/kg.
- Gravitational attraction also depends on the distance between the two objects. The larger the distance between two objects, the weaker the gravitational attraction.
- To escape from the Earth, rockets need a **thrust** greater than their weight.
- As a rocket travels away from the Earth, the gravitational attraction on it decreases because its distance from Earth gets bigger.

J2 Satellites

Learn about:

- Artificial and natural satellites
- Staying in orbit

Sputnik

Any object that orbits a larger object is called a **satellite**. In 1957 Sputnik 1 orbited the Earth for 52 days. It was the first **artificial satellite**, a satellite made by people. Sputnik 1 (shown in this photo) had a mass of 83.6 kg and a diameter of 58 cm. It orbited the Earth every 96 minutes. It was put into orbit by the Soviet Union to collect information that they later used to put people into space.

Natural satellites

We usually use the word 'satellite' to mean artificial satellites, but any object that orbits a bigger object is a satellite. The Earth is a **natural satellite** of the Sun and the Moon is a natural satellite of Earth. In general, moons are satellites of planets and planets are satellites of stars.

Did you know?

The word 'sputnik' means travelling companion in Russian.

Staying in orbit

To stay in orbit, a satellite has to be moving fast. There is gravitational attraction between it and Earth, pulling it down. A satellite is constantly falling, but the trick is to get it to fall around the Earth rather than down to Earth.

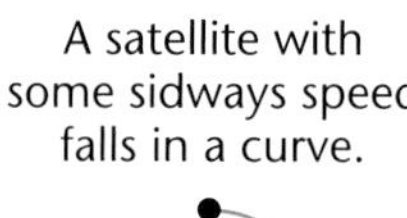

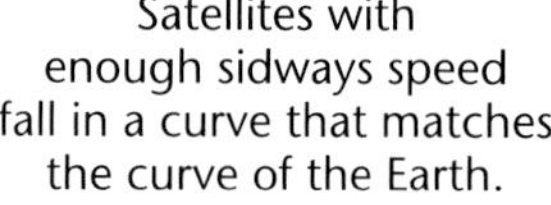

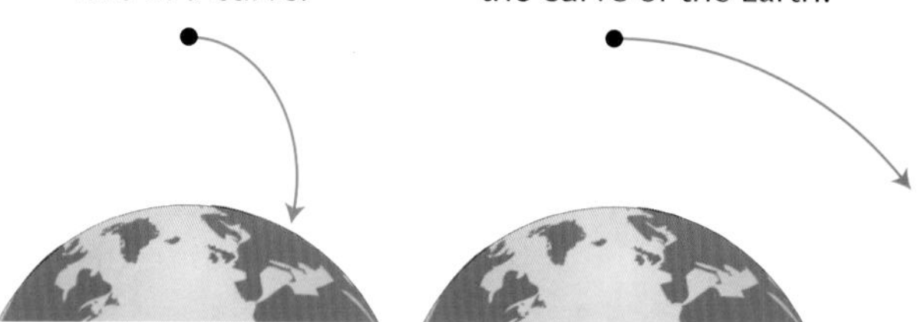

It seems odd that a satellite stays at the same height and travelling at the same speed when there is a large force, gravitational attraction, pulling it down to Earth. Look at this diagram on the right. The green arrows show the direction the satellite is travelling in at five points in its orbit. What is the force of gravitational attraction doing? The answer is that it is making the satellite change direction all the time, so that it travels in a curve (the red line) rather than a straight line.

If you could switch off gravitational attraction, then the satellite would fly away in the direction shown by the green arrows. The speed of the satellite has to be just right. Too slow and it will crash into the Earth. Too fast and it will fly off into space. Sputnik 1 fell back to Earth after 52 days because it started to slow down.

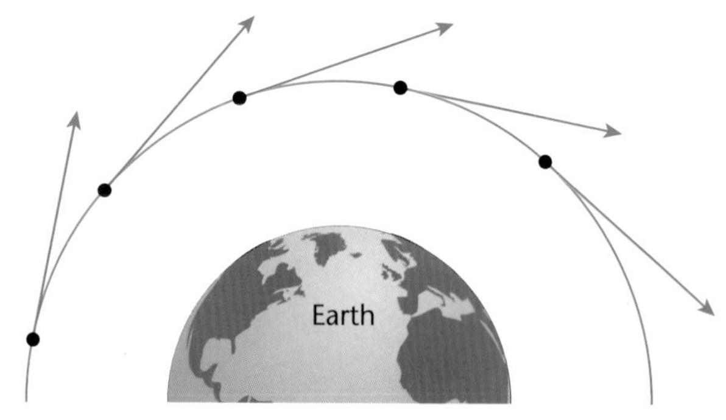

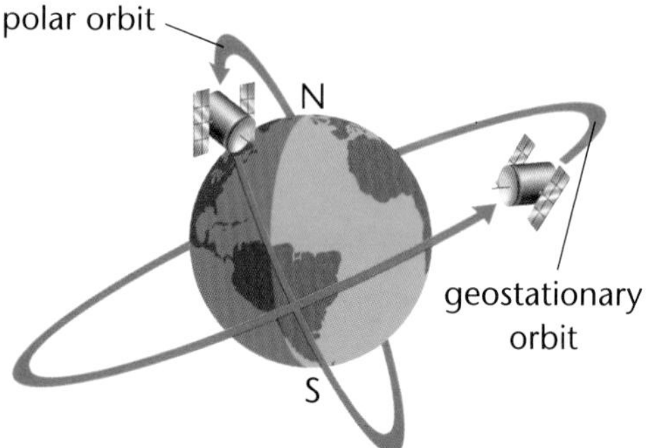

Different orbits

Artificial satellites are put into different orbits depending on the job they are going to do. Some orbit the Earth at the same speed as the Earth is turning on its axis. These satellites are in **geostationary orbits**. Geostationary satellites orbit the Earth once every 24 hours, which means they stay in the same place over the Earth's surface. Other satellites are in orbits that take them over the North and South Poles. These are called **polar orbits**.

Uses of artificial satellites

Artificial satellites are useful to us in many ways.

Type of satellite	Function
communication satellites	Send radio, TV and telephone messages around the world. These are geostationary satellites.
exploration satellites	Carry telescopes and take clear pictures of planets. Can also look at the universe. The Hubble telescope is an example.
navigation satellites	Used by ships, cars and planes to find their position on the Earth to within a few metres.
observation satellites	Take detailed photos of the Earth. Can show volcanic eruptions, floods and oil spills. Also used to spy on other countries. Can take pictures of cloud formations and send them to Earth to be analysed to help weather forecasters to predict our weather.
space stations	Astronauts and cosmonauts live and work in these. Used to conduct scientific experiments, maintain other satellites and collect information about living in space. The ISS, or International Space Station, has been constantly crewed since November 2000.

The International Space Station.

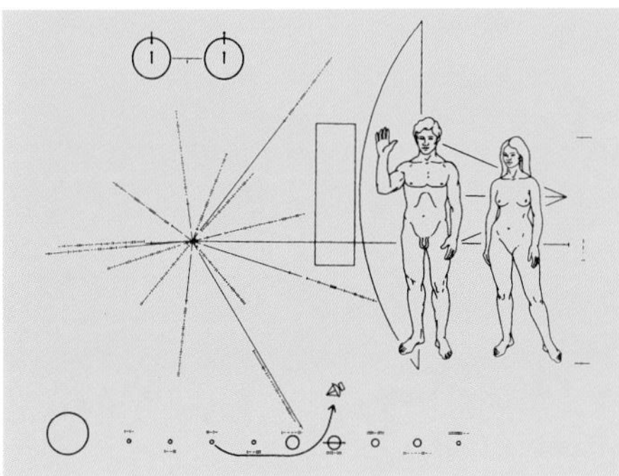

Pioneer 10 carried this message about humans and Earth into space.

Exploring space

We have also sent many spacecraft out into space. The famous Apollo missions took humans to the Moon, but uncrewed spacecraft have gone much further. Some have landed on Mars, while others have sent back images of the Solar System. Pioneer 10, launched in 1972, sent its last signal to Earth in 2002. It has now run out of power, but at last count it was 12 billion kilometres from Earth, carrying humanity's message into deep space.

Questions

1 **a** What force keeps satellites in orbit?

b What happens to a satellite that is travelling too slowly?

c What happens to a satellite that is travelling too fast?

2 What is the advantage of a satellite that is in:

a a geostationary orbit? **b** a polar orbit?

3 People are launching more and more satellites into space. Why could this cause problems in the future?

4 The Russians planned to use a giant mirror in space to reflect light from the Sun onto a city at night.

a What are the advantages and disadvantages of this idea?

b Draw a diagram to show how this would work.

For your notes:

- A **satellite** is an object that orbits a larger object.
- The speed of a satellite has to be just right so that the gravitational attraction keeps it in orbit.
- **Artificial satellites** are machines launched into space by people. They are used for communicating across our planet or studying space.
- There are **natural satellites**, such as the Earth and other planets orbiting the Sun, and the Moon orbiting the Earth.

J3 The Solar System

Learn about:

- Models of the Solar System

Pancake Earth?

You may have heard that in ancient times everyone thought that the Earth was flat. But that isn't true – the idea that Earth was a sphere was widely accepted among the great thinkers of Ancient Greece, including Thales and Pythagorus. Aristole (384–322 BC) wrote down the arguments that supported the idea that Earth was a sphere, and this became the view accepted by most scholars.

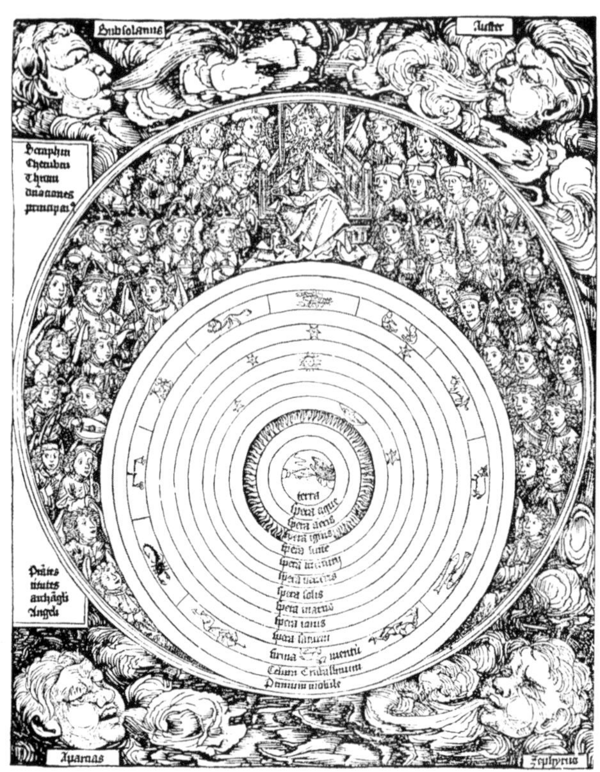

Earth at the centre?

A number of Greek philosophers, including Aristotle, thought the Earth was the centre of the universe and that everything including the Sun moved around the Earth. This model shown in this photo was called the **geocentric model** of the universe. 'Geo' comes from the Greek word for Earth. For over 1000 years this theory was not challenged.

Did you know?

There was one Ancient Greek thinker, Aristarchus, who thought that the Earth went round the Sun. He lived from 320 BC to 250 BC. Unfortunately, the idea did not catch on for a long time.

Sun at the centre?

The Polish astronomer Nicolaus Copernicus (1473–1543) suggested that the Sun was at the centre of the Solar System, with the planets going around the Sun in circular orbits. This was called the **heliocentric model**. 'Helio' is from the Greek for Sun. He based his model on observations he made over 30 years without the aid of a telescope, which had not been invented. The model Copernicus drew is shown on the right.

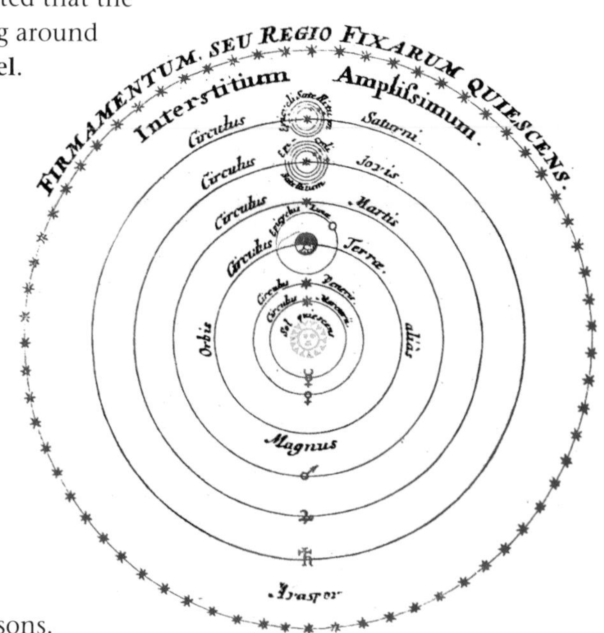

If astronomers had been able to look down on the universe and see the Sun, Earth and Moon, they would have seen exactly how they all moved. Instead, they had to make all the observations they could and use these as evidence for their models. They would have considered things like:

- the apparent size of the Sun, Moon and stars
- the regular changes of day and night and of the Moon's shape
- how the length of the day changes with the time of year
- how the position of the Sun in the sky changes with the seasons.

a **What might the apparent size of the Sun and the Moon tell you about their distances from the Earth?**

b **What do all the regular changes such as day and night and the seasons suggest to you?**

The Italian astronomer Galileo Galilei (1564–1642) designed and made telescopes. These allowed him to see things in space magnified 30 times. He discovered the moons of Jupiter and Saturn's rings. He observed that Jupiter's moons appeared move around Jupiter and this convinced him that the heliocentric theory of Copernicus was correct.

Galileo demonstrating his telescope.

A dangerous idea

In Europe at that time, the Roman Catholic Church was very powerful. It was a crime to openly disagree with the Church's teaching that the geocentric model was 'right'. The punishment was death. Copernicus lived in Poland and only published his ideas when he was an old man. Galileo lived and published his work in Italy, where the Catholic Church had its headquarters. In 1633, Galileo was put on trial for supporting the heliocentric model. He had to make a public statement that he was wrong, and managed to avoid execution. Instead he was imprisoned for the rest of his life.

Refining the model

Tycho Brahe (1546–1601) was a Danish astronomer who made very accurate star charts. These included the positions of the planets, especially Mars.

The German mathematician Johannes Kepler (1571–1630) was an assistant to Tycho Brahe. Kepler used Brahe's observations to write his laws of planetary motion. He showed that the planets did not move in perfect circles. Instead the orbits were ellipses, or flattened circles. Scientists used Kepler's calculations to make excellent predictions of how planets move. The only problem was, Kepler did not understand why his laws worked!

The British scientist Isaac Newton (1642–1727) explained Kepler's laws in 1687. Newton worked out two relationships. The bigger the masses of the two objects, the larger the gravitational attraction between them. The further the objects are apart, as measured from their centres, the smaller the gravitational attraction between them.

Tycho Brahe's observatory.

C **A 1 kg mass on the surface of Saturn weighs about the same as it does on Earth, almost 10 N. Saturn has a mass 95 times greater than the mass of the Earth. Saturn has a radius that is 10 times greater than the radius of the Earth.**

(i) Think only about the masses: where would expect the 1 kg mass to weigh more, on Earth or on Saturn? Explain your answer.

(ii) Think about the distance between the centre of the 1 kg mass and the centre of the planet. Explain why the 1 kg mass weighs the same on Saturn and the Earth.

Questions

1 Imagine you live in a tribe in prehistoric times and you know nothing about the Solar System. Your territory is large and flat but surrounded by impenetrable forests. Write down some of the observations (not explanations) you would be able to make about the Earth, Moon and Sun.

2 Explain why data and observations are important to people who are developing theories and models.

3 Draw a flow chart to show how each scientist's model of the Solar System was developed from the earlier models of others.

For your notes:

- Many models have been put forward over the centuries to explain the Solar System.
- These models have been changed and improved as new observations and evidence became available.
- Copernicus first put forward the **heliocentric model**, with the Sun at the centre of the Solar System, in the fifteenth century.

Think about:

- Theories and evidence

J4 Birth of the Moon

An odd Moon

Moons orbit planets. Mars has two moons. At the latest count, Jupiter has 50. Most moons are small compared with the planet they orbit. Our Moon is different. It is huge! The Moon's diameter is one-quarter the diameter of the Earth. How did Earth end up with such a huge satellite?

How did it form?

There were three scientific ideas, or theories, of how the Moon formed.

1 *The spin theory*
In 1878 George Darwin suggested that that the Moon was originally part of the Earth. When the Earth was young it was very hot and spun very fast. The surface layers of rock were molten, like hot toffee. Darwin suggested a lump was pulled off by the Sun's gravitational attraction.

The problem with Darwin's idea is that it needs the Earth to be spinning very fast. Scientists knew that the Earth spun faster than they expected for a planet of its size and position. However, there is no evidence that it ever spun fast enough to spin off a Moon-sized lump.

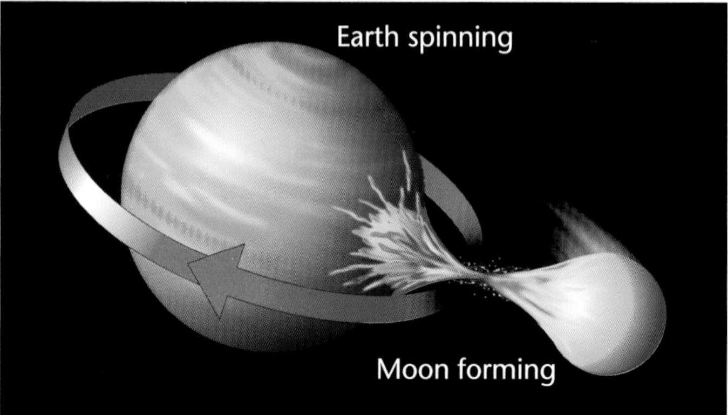

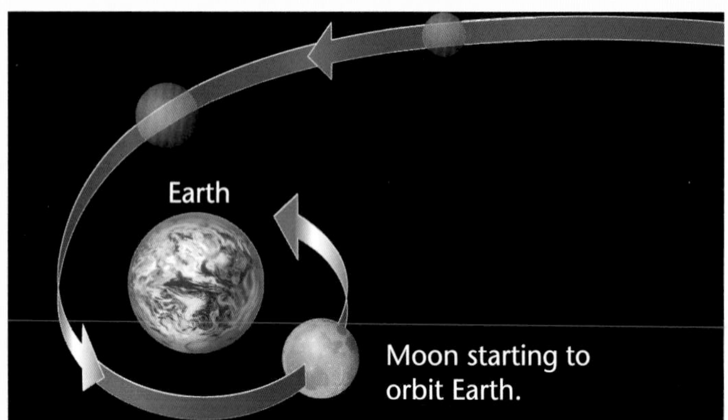

2 *The capture theory*
In 1909 Thomas See suggested that the Moon was 'caught' by Earth's gravitational attraction. In See's theory the Moon formed somewhere out in space. It passed too close to the Earth and was captured.

The problem with See's theory is that the large Moon would have to pass close enough to the Earth to be caught, but not to hit the surface or miss entirely. Scientists did not think this very likely.

3 *The double planet theory*
Most scientists supported the idea that the Earth and Moon formed at the same time. Earth and Moon were a 'double planet'. This explained why the Moon was so large.

The problem with this theory was that the Moon is one-quarter the size of the Earth but much less than one-quarter of its mass. The density of the Moon is much lower than the density of the Earth. This seems a bit odd if they formed at the same time.

a **Which theory suggests the Moon formed from the surface of the Earth?**

b **Which theory suggests the Moon formed in exactly the same way as the Earth formed?**

c **Do you expect rock from the Earth and rock from the Moon to be the same? Use the three theories to suggest three possible answers.**

Collecting evidence

Scientists had based their theories on their observations of the Moon from Earth and on models about how the planets formed. Scientists needed more evidence. In the late 1960s and early 1970s, astronauts and robot spacecraft collected about 400 kg of Moon rocks. Analysing the Moon rocks gave the scientists evidence that the Moon had once been part of the Earth, or had formed very close by.

d Which theory does not fit with this new evidence?

The Moon rocks did not contain any iron. This made the 'double planet' theory very unlikely. As one scientist, William Hartmann, said, 'It's very hard to imagine the two bodies growing together but somehow the Earth magically gets all the stuff with the iron in it and the Moon doesn't get any.'

Suddenly the 100-year-old spin theory seemed most likely to be right. The theory said that the Moon was made from only the outer layers of the Earth. These layers don't contain much iron.

But there were two problems with the spin theory. First, there was no evidence that the Earth has ever spun fast enough to spin off a Moon-sized lump. Second, the Moon rocks had very few **volatile** compounds, fewer than the outer layers of the Earth. Volatile substances boil away when a mixture is heated. The spin theory did not have an explanation for this.

The giant impact theory

In 1975 William Hartmann and Donald Davis proposed a new theory. They suggested that Earth had been hit by an object, the size of a small planet. This had happened about 4.5 billion years ago, after the Earth's iron core had formed. Material from the outer surface of the Earth was blasted into orbit. It also caused huge temperature rises so the more volatile substances boiled away into space. The remaining material from the Earth slowly came together to form the Moon.

e How does the giant impact theory explain the following evidence?
- **(i) The Moon contains almost no iron.**
- **(ii) The rocks on the Moon are very similar in many ways to the rocks in the outer surface of the Earth.**
- **(iii) Moon rocks contain fewer volatile substances than Earth rocks.**

Other scientists were also working on the same idea. In 1976 A.G.W. Cameron and William Ward came to the same conclusion based on a completely different approach. They looked at how quickly the Earth and Moon spin, which is faster than expected. Cameron and Ward suggested that the extra spin came from the impact of an object about the size of Mars.

Since 1975 many scientists have worked on the giant impact theory. There are computer models that show what might have happened. The giant impact theory is now accepted by the majority of scientists.

Did you know?

Astronauts first walked on the Moon on 15 July 1969.

Collecting rock from the Moon.

Giant impact theory.

Questions

1 What types of evidence could scientists collect about the Moon before 1969?

2 What changed in 1969?

3 How did the evidence contained in the Moon rocks change the ways scientists thought about the origin of the Moon?

4 Why was it important that Cameron and Ward came to the same conclusion as Hartmann and Davis?

K1 Racing

Learn about:

- Speed
- Acceleration

Distance and time

The athletes in the photo below are competing in a 100 m race. The athlete with the shortest time wins. So speed depends on time. The shorter the time taken, the faster the speed.

$$\text{speed in metres per second} = \frac{\text{distance travelled in metres}}{\text{time taken in seconds}}$$

Do you remember?

We use distance and time to work out speed.

Did you know?

Athletes are running faster! For example, the world record for the 400 m has decreased over the years. In 2000 it was 43.2 s. In 1970 it was 44.4 s, in 1940 it was 46.0 s and in 1900 it was 47.8 s.

The winning 100 m athlete, Tim Montgomery, ran 100 metres in 9.78 seconds.

$$\text{Tim's speed} = \frac{100 \text{ metres}}{9.78 \text{ seconds}}$$
$$= 10.2 \text{ metres per second or } 10.2\,\text{m/s}$$

It is very important to include the **units** when you are talking about speed. Tim Montgomery's speed was 10.2 m/s. If you say just '10.2' you could mean 10.2 miles per hour, which is a slow jog rather than a record-breaking sprint.

Ray and his friends are racing snails. The winning snail is the one that crawls the furthest along the table in 5 minutes. Speed depends on distance. The longer the distance travelled, the faster the speed.

a **Lightning, the winning snail, crawls 1.4 metres in 5 minutes.**
(i) What is Lightning's speed?
(ii) How much faster is Tim Montgomery than Lightning?

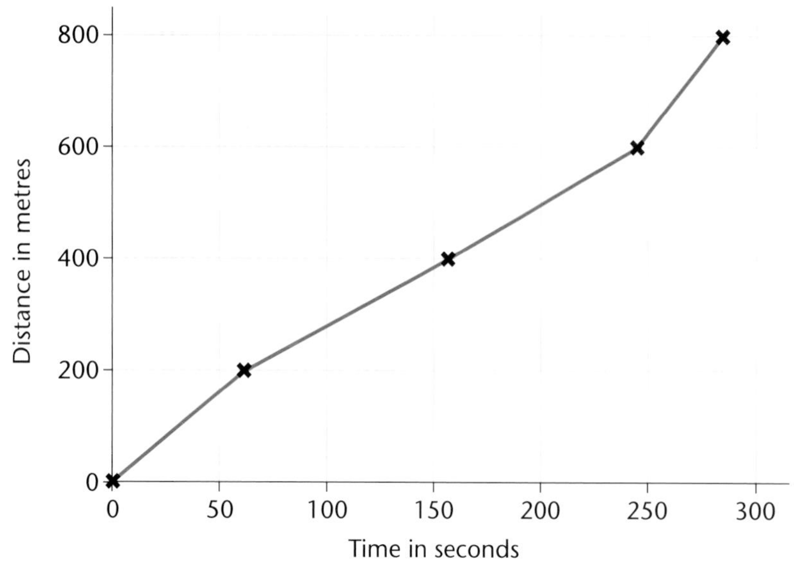

Showing speed

Susan is running an 800 m race. Her coach has four stopwatches and stops one after 200 metres, 400 metres, 600 metres and 800 metres. This graph shows the times for the four distances.

Look at the **distance–time** graph. The slope or **gradient** of the graph shows the speed. The steeper the gradient, the faster the speed. You can see that Susan runs most quickly for the last 200 metres.

The coach's assistant videos Susan running. He uses the video to work out her speed every 10 seconds. He then draws the **speed–time graph** opposite.

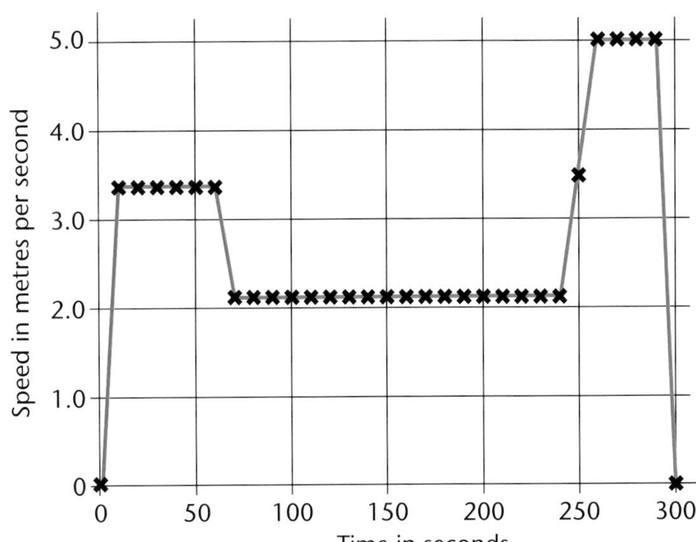

On a speed–time graph, steady speed is shown by a horizontal line. Susan runs at three different steady speeds: 3.3 m/s, 2.1 m/s and 5.0 m/s.

When Susan speeds up, the graph gradient is upwards. We say that Susan **accelerates**. Speeding up is **acceleration**. When Susan slows down, the graph gradient is downwards. We say that Susan **decelerates**. Slowing down is **deceleration**.

b **To measure speed you need to know the distance and time. Explain how the coach could set up the running track and video to help him work out Susan's speed every 10 seconds from the video recording.**

c **How many periods of:**
(i) **acceleration and**
(ii) **deceleration are shown on the speed–time graph?**

Questions

1 Harriet cycles 220 metres in 22 seconds at a steady speed.

a What is her speed?

b How far would she cycle in 10 seconds?

c How long would it take her to cycle 400 metres?

2 Sean is a cyclist. He takes part in a 20 km cycle race. As well as the overall prize for winner, there are prizes for three sprints during the race. This distance–time graph shows Sean's race.

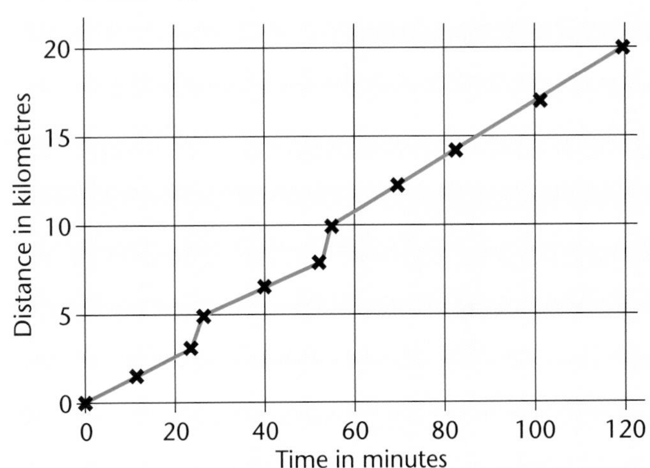

a How many of the sprints did Sean take part in? Explain your answer.

b Ignoring the sprints, what was Sean's overall speed for the race in: **(i)** km/min? **(ii)** km/h?

Look at these three speed–time graphs.

c Which of these speed–time graphs shows Sean's race?

d Which of them shows a cyclist who took part in only one sprint?

e Which of them shows a cyclist who had a puncture?

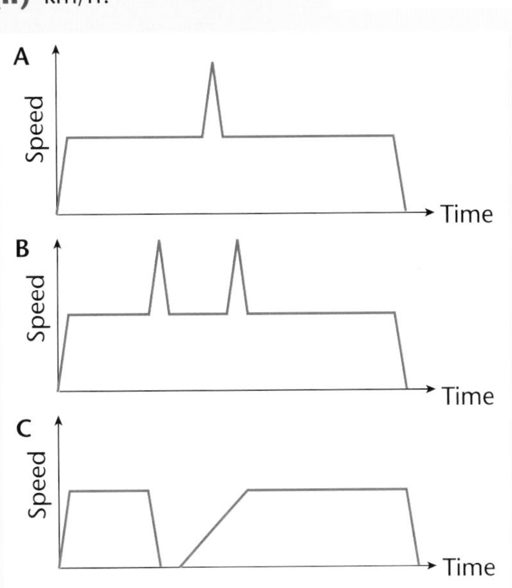

For your notes:

- Speed is distance travelled ÷ time taken. The **unit** for speed is m/s or km/h.
- Steady speed is shown as a **gradient** on a **distance–time graph** and as a horizontal line on a **speed–time graph**.
- **Acceleration** is an upwards gradient on a speed–time graph.
- **Deceleration** is a downwards gradient on a speed–time graph.

K2 Measuring speed

Learn about:

- Measuring instruments
- Precision, reliability, accuracy

Measuring distance and time

Each pair of pupils in Janet's class is measuring the speed of sound. Janet and Luisa decide they are going to measure the time it takes a sound to travel the length of the school field.

The first thing they have to do is choose the measuring instruments they are going to use. Three instruments used to measure distance are shown in this photo.

Each of the instruments has a different **precision**. The measuring wheel measures to the nearest 1 metre. The tape measure measures to the nearest 1 centimetre, the ruler to the nearest 1 millimetre. The ruler is more **precise** than the measuring wheel. Luisa chooses the measuring wheel to measure the school field.

a **Do you agree with Luisa's choice? Explain your answer.**

Janet chooses a digital stopwatch to measure the time. Sound travels very quickly. It will cross the field in a very short time. The watch needs to be very precise. The digital stopwatch shows time to 0.01 seconds.

How fast is sound?

Janet goes to the other end of the playing field. She measures the distance with a measuring wheel. She ends up 300 m away from Luisa. Luisa has a large wooden clapper. Janet starts her stopwatch when she sees Luisa close the clapper, and stops the stopwatch when the sound reaches her. Janet measures the time as 0.95 s.

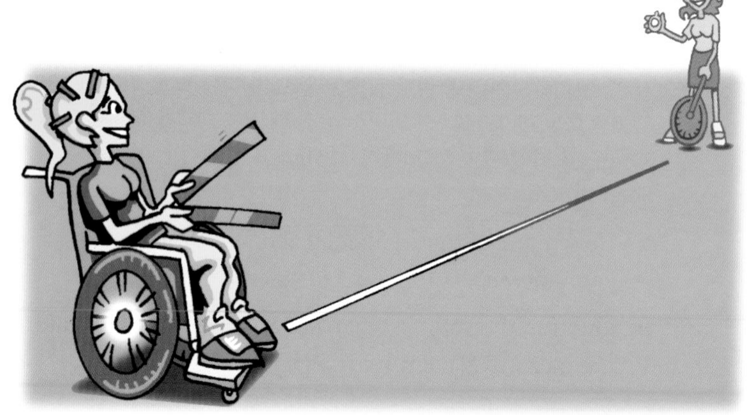

Luisa and Janet calculate the speed of sound as 316 m/s.

b **Do your own calculation to confirm that Janet and Luisa are correct.**

They tell their teacher, Mrs Brook, that the speed of sound is 316 m/s.

How reliable, how accurate?

Mrs Brook asks Janet and Luisa how **reliable** their value is for the speed of sound. A value becomes more reliable the more times it is measured. Its **reliability** improves.

The two girls are not sure and decide to repeat the experiment five more times, calculating the speed each time. Their results are shown here.

Experiment	1	2	3	4	5	6
Time in s	0.95	0.94	0.88	0.92	0.86	0.90
Speed in m/s	316	319	341	326	349	333

Not all their values for the speed of sound are the same. They decide to calculate a mean (an average) of their results.

$$\text{mean} = \frac{\text{sum of the values}}{\text{number of values}}$$

Janet and Luisa's mean is 331 m/s.

c **Check Janet and Luisa's calculation by doing your own.**

The value for the speed of sound given in books is 330 m/s. By repeating the experiment and working out the mean, Janet and Luisa found out a more **accurate** value for the speed of sound. The closer a value is to the 'real' value, the higher its **accuracy**.

You can also make a value more accurate by taking the measurements very carefully and using more precise measuring instruments.

Speed of light

Sound travels quickly. The problem with the experiment is measuring the time taken, which is very small. Janet and Luisa made the distance as large as possible and used a very precise stopwatch.

Imagine how much more difficult it is to measure the speed of light! A scientist called Albert Michelson set out to do just that in 1882. He needed a very big distance, so he chose two mountain peaks that were 36 kilometres apart. No clock or watch was precise enough, so he built an ingenious timing device with a spinning, octagonal mirror that could measure time to 0.000 01 seconds.

d **Albert Michelson measured the time taken for light to go from one mountain peak to the other and then back again. The time taken was 0.000 24 s.**

(i) What distance did the light travel?

(ii) What was the speed of the light?

Measuring speed

When we measure speed, it is sometimes not obvious we are measuring both distance and time. A car speedometer measures both the distance travelled and the time taken. Light gates, like the ones in this photo, can measure speed because there is a clock built into the computer to measure time, and we tell the computer the distance between the light gates.

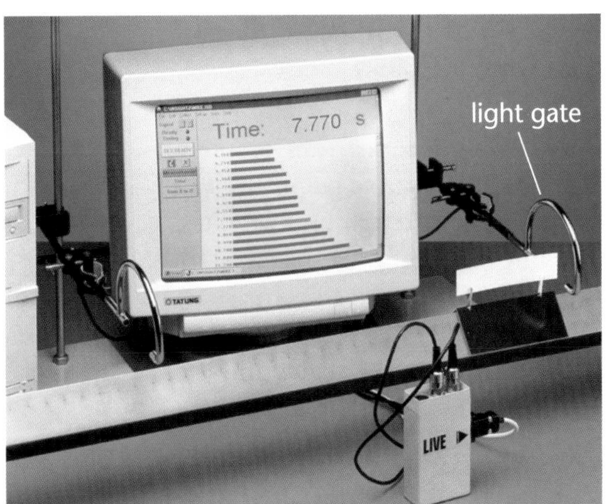

Questions

1 Look at these times. They are the same time measured differently. Put them in order with the most precise first and the least precise last.

84 706 seconds **a day** **1412 minutes** **23.5 hours**

2 Four pupils have to measure the length of a stick as part of a science challenge. The stick is exactly 832.72 mm long. Which is the most accurate measurement?

A 832.02 mm **B** 831.56 mm **C** 832.10 mm **D** 833.12 mm

3 Kevin and Malcolm are measuring the speed of sound using echoes. They stand 20 metres from a tall brick wall. They blow a whistle and use a stopwatch (accurate to 0.01 seconds) to measure the time it takes for the sound to reach the wall and come back to them.

a It takes the sound 0.13 seconds to travel to the wall and back. What is their value for the speed of sound?

b Explain how these modifications would improve their experiment:

(i) repeating the experiment 10 times
(ii) using a stopwatch that measured to 0.001 seconds
(iii) standing 40 metres from the wall
(iv) using a starting pistol with the trigger linked to an automatic timer.

For your notes:

- A more **precise** measuring instrument measures in smaller amounts, for example millimetres instead of centimetres.
- A more **reliable** value comes from repeating a measurement and taking a **mean** (an average).
- A more **accurate** value is closer to the real value.

K3 Changing speed

Learn about:

- Steady speed
- Acceleration
- Deceleration

Steady speed

John plays ice hockey. He is taking part in a science experiment about forces and speed. John skates around the rink and then glides 10 metres. There are light gates every 2 metres along the glide path. John breaks the beam when he glides past them and a computer records the time.

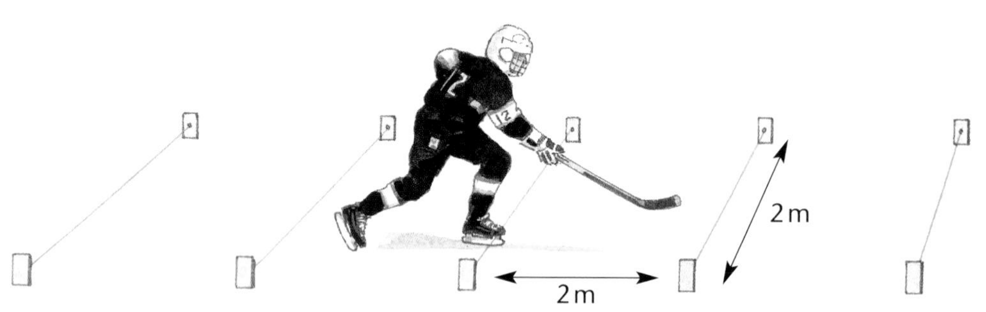

Look at this distance–time graph. The **blue** line on the graph shows John gliding through on his skates. The line is straight, with a constant gradient. John is travelling at a **steady speed**. There is almost no friction between the skates and the ice, so he does not slow down.

a **How would the blue line be different if John skated at a higher steady speed?**

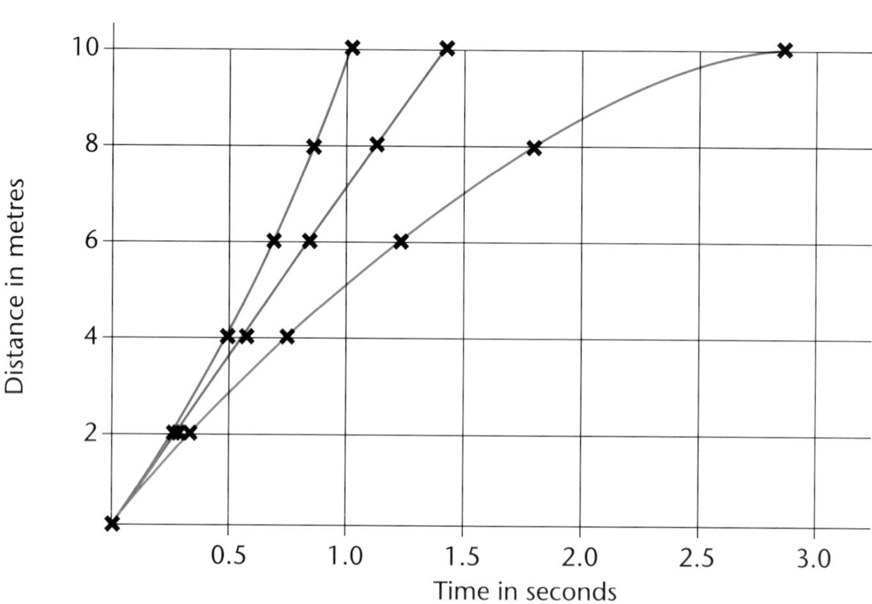

resultant force

friction

push

Speeding up

John then skates through the light gates, pushing as hard as he can against the ice. The **brown** line on the graph above shows the results. The graph is not straight because John is not travelling at a steady speed. The graph curves upwards: the gradient of the graph is increasing. John is accelerating.

John accelerates because he is pushing against the ice. The push forward is a lot larger than the friction. The forces are unbalanced and the forward force is larger.

The blue arrow in the diagram on the left shows the resultant force. The resultant force is the overall force. If there is a 100 N force forward and a 10 N force backwards, then there is an overall or resultant force of 90 N in the forwards direction. The resultant force causes John to speed up.

b **On another day the ice is less smooth. John still pushes with a force of 100 N but the friction is 25 N. Compare John's acceleration on the two days.**

John repeats the experiment, but with a heavy rucksack. The rucksack increases his mass. He still manages to speed up, but it is more difficult, and he accelerates less. It takes more force to speed up objects with more mass. John is already pushing with all the force he can manage, so he accelerates less.

c Use this information to suggest why lorries need more powerful engines than cars.

Slowing down

Then John goes through the light gates on his knees. His ice hockey kit will protect his knees.

The **pink** line on the graph on page 102 shows John sliding through on his knees. The line is not straight. The graph curves downwards: the gradient of the line is decreasing. John is slowing down. The forces are unbalanced. There is a force backwards, friction, but no force forwards to balance it. The unbalanced force slows John down.

d What would make John slow down even more? Give a reason for your answer.

John repeats the experiment wearing the rucksack. He slows down less. The friction between John and the ice is the same, so the same force is slowing down a larger mass. The larger the mass, the harder it is to slow down, just like it was harder to speed up.

e Does a lorry or a car need more force to slow it down? Explain your answer.

Do you remember?

Forces are measured in newtons, N. We show forces using arrows. The length of the arrow shows the size and direction of the force.

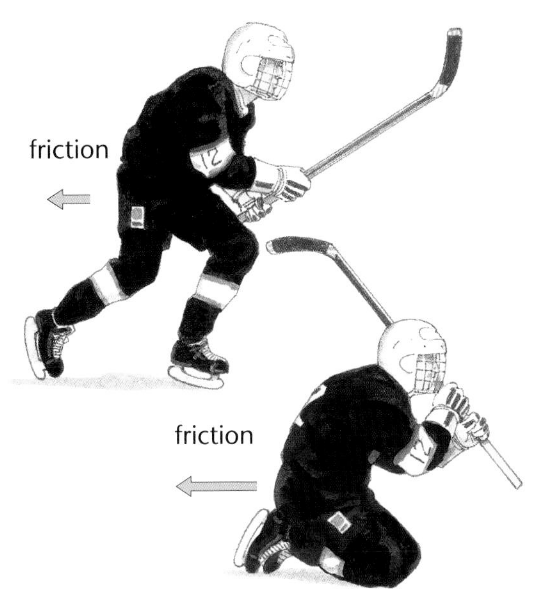

Questions

1 A skater pushes forward with a force of 30 N. The friction is 5 N.

- **a** What is the resultant force?
- **b** Is the resultant force forwards or backwards?
- **c** Does the skater speed up, slow down, or travel at a steady speed?

2 Ellen skateboards to her friend's house. Her journey is shown on this distance–time graph.

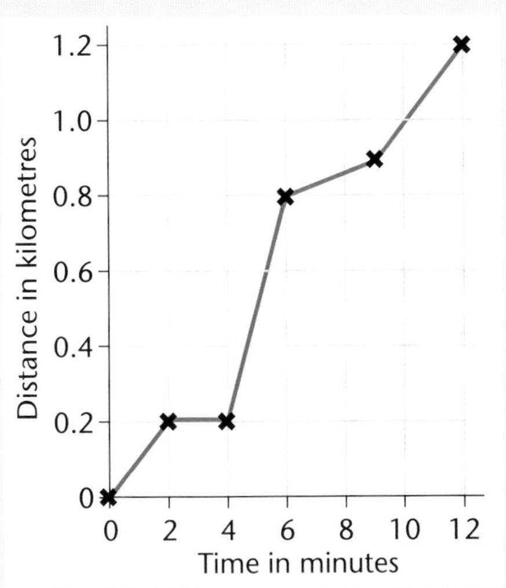

- **a** Between which two times does Ellen skateboard most quickly?
- **b** Between which two times does Ellen stop?
- **c** Calculate the *overall* speed for the whole journey.
- **d** Sketch a speed–time graph of the same journey.

For your notes:

- If the forces are unbalanced, you can work out the size and the direction of the resultant force. If the forces are balanced there is no resultant force.

Resultant force	Movement	Distance–time graph
none	steady speed	straight line – gradient shows speed
forwards	accelerating	upwards curve with increasing gradient
backwards	decelerating	downwards curve with decreasing gradient

- The larger the mass, the larger the force you need to speed it up or slow it down.

K4 Faster!

Learn about:

- Thrust
- Streamlining

Thrust SSC

The world land speed record was broken in 1997 by Andy Green driving *Thrust SSC*, shown in this photo. *Thrust SSC* travelled at a mean (average) speed of 763.04 mph.

A force pushes the car forward. This force is produced by the car's engines. It is called **thrust**, which gave the car its name. *Thrust SSC* had two very large jet engines (the type used on planes). Each engine can produce a force of 110 000 N, or 110 kN. That is the same as the thrust produced by 500 family cars.

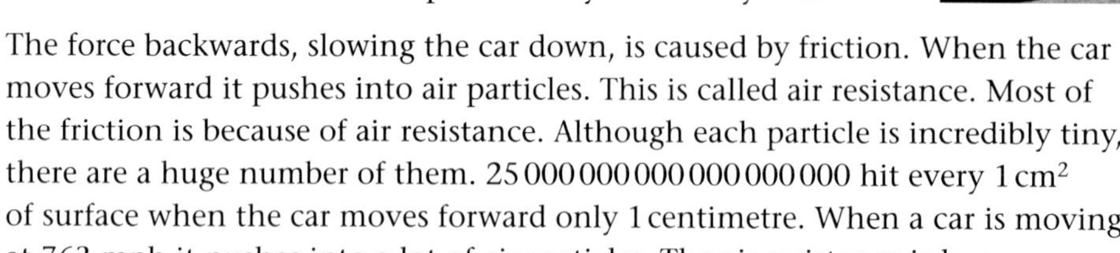

The force backwards, slowing the car down, is caused by friction. When the car moves forward it pushes into air particles. This is called air resistance. Most of the friction is because of air resistance. Although each particle is incredibly tiny, there are a huge number of them. 25 000 000 000 000 000 000 hit every 1 cm^2 of surface when the car moves forward only 1 centimetre. When a car is moving at 763 mph it pushes into a lot of air particles. The air resistance is huge.

Maximum speed

When the car starts up it is moving slowly. It is pushing into fewer air particles every second than later on, so the air resistance is low. The thrust is much larger than the air resistance. The forces are unbalanced and the resultant force makes the car speed up. This is shown in the left-hand diagram here.

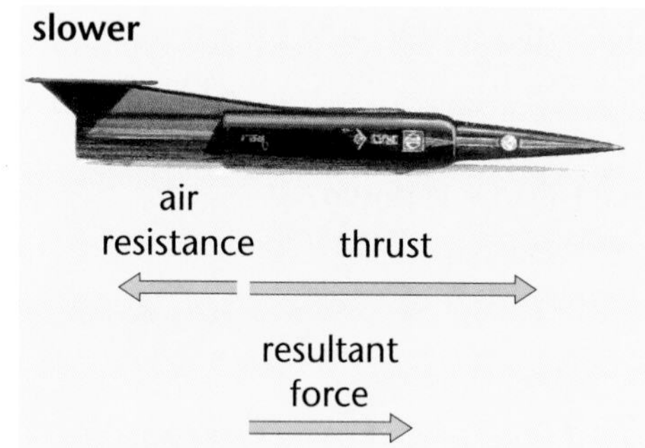

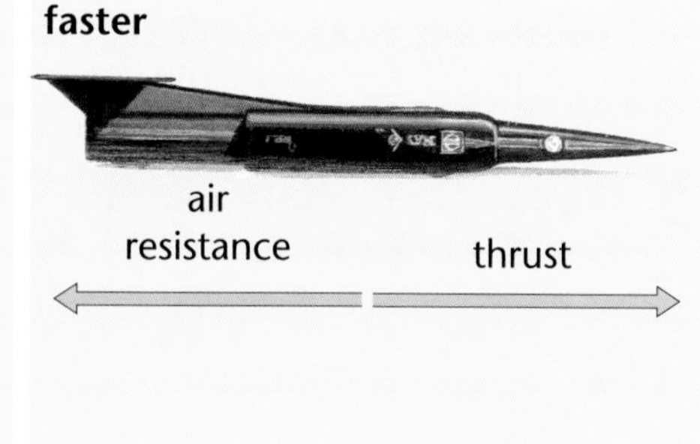

As the car speeds up it pushes into even more air particles. The air resistance increases. Finally, the air resistance balances the thrust. The car now goes at a steady speed. This is its maximum speed. This is shown above on the right.

a **Explain why the air resistance is greater for a fast-moving object than for a slow-moving object.**

b **Why does the *Thrust SSC* car reach a maximum speed?**

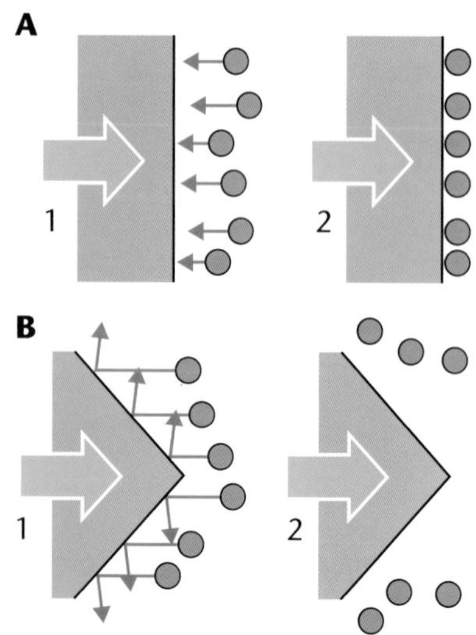

Streamlining

Cutting down the air resistance increases the maximum speed. It means the car can go faster before the air resistance balances the thrust at its maximum speed.

Smooth, sleek shapes have lower air resistance and are more **streamlined** than lumpy, boxy shapes.

Diagram A shows what happens when you push a flat surface through air. The red balls represent a few of the particles in the air. The surface hits the particles at right angles. The particles will end up being pushed forward, in front of the surface. The engines end up pushing the air, as well as the car.

Diagram B shows what happens with a more streamlined shape. The surfaces are at an angle. The particles bounce sideways. The air ends up to the sides, rather than in front. The engines do not have to push the air as well as the car.

The designer of *Thrust SSC* made it as streamlined as possible. Every surface curves. No part is at right angles to the way the car will be going.

The friction between a moving object and water is often called **drag**. Fast-moving objects are often streamlined to reduce drag or air resistance. Aeroplanes push past many air particles when they fly. Boats and submarines push past many water particles.

Even small bumps on the surface can increase the air resistance or the drag. That's why Olympic swimmers shave off all the hair on their bodies and wear a tight-fitting cap on their heads. Some modern swimmers even wear special body suits like in this photo because the material of the suit causes less drag than skin.

c Penguins are very streamlined. What survival advantage does this give them?

Fuel efficient, not fastest

Thrust SSC may be the fastest land vehicle, but it isn't the most fuel efficient. Every year a competition is held to find the most fuel-efficient design of car. The winner is the car that can travel the furthest on one gallon of fuel. Qualifying cars must have a driver, three or four wheels, travel at an average of 15 mph and complete 10 miles of the course. The winning car in 1996 managed 568 miles per gallon! One of the competing cars is shown in the photo on the left.

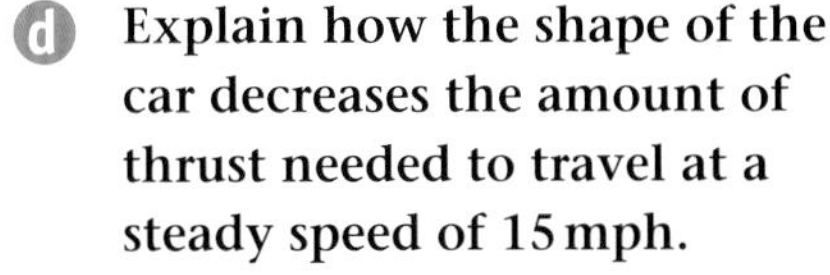

d Explain how the shape of the car decreases the amount of thrust needed to travel at a steady speed of 15 mph.

e Why does this lead to a smaller petrol consumption?

Questions

1 The table describes the forces on a car.

	Thrust in N	Friction (including air resistance) in N	Resultant force		Speeding up, slowing down, or steady speed
			Size in N	Direction	
A	200	10			
B	100	150			
C	200	200			

a Copy the table and complete the last three columns.

A car goes down a sliproad onto a motorway, travels along the motorway at high speed and then brakes because there is an increase in traffic.

b Put A, B and C from the table into the correct order to fit with the car's movement.

2 Look at the photo below of a hydrofoil. A hydrofoil is a vehicle that travels along the surface of the water. Use your knowledge and understanding of particle theory to explain:

a why it takes less thrust to push a hydrofoil through air than a boat through water

b why the streamlined 'boat' shape gives the hydrofoil a higher maximum speed than a more 'boxy' shape.

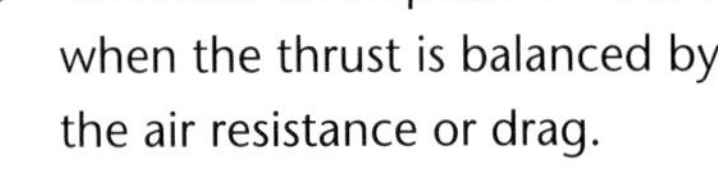

For your notes:

- The pushing force of an engine is called **thrust**.
- The friction of a vehicle pushing past particles is called air resistance or **drag**.
- The air resistance or drag is larger when the vehicle is travelling faster, because it pushes past more particles.
- The maximum speed is reached when the thrust is balanced by the air resistance or drag.

K5 Slow down

Learn about:

- Parachutes
- Brakes

Stop!

This photo shows a drag racer. Drag racers have engines that give huge thrust, but no brakes. When they cross the finishing line they turn off the engine, but they only have air resistance to slow them down.

The parachute increases the air resistance because it 'catches' the air. The air particles hit the parachute, slowing the drag car down.

a **Why does a drag car racer want to minimise air resistance during the race but maximise it when the race is over?**

Falling

A skydiver (with equipment) has a mass of 100 kg. This means that her weight is 1000 N. The weight always acts downwards, towards the centre of the Earth. Air resistance acts upwards. That is because it is in the opposite direction to the movement, and she is falling.

The size of the air resistance depends on the speed of her fall and whether or not she has opened the parachute. The faster she is falling, the greater the air resistance, because she is pushing past more particles every second. An open parachute has a huge surface for the air particles to hit, increasing her air resistance dramatically.

	Event	Weight in N	Air resistance in N	Resultant force in N (up or down)	Accelerating, declerating or steady speed?
1	exits plane	1000	0	1000 down	accelerating
2		1000	600		
3	free fall	1000	1000		
4	parachute fully open	1000	2000		
5		1000	1600		
6	safe speed achieved	1000	1000		

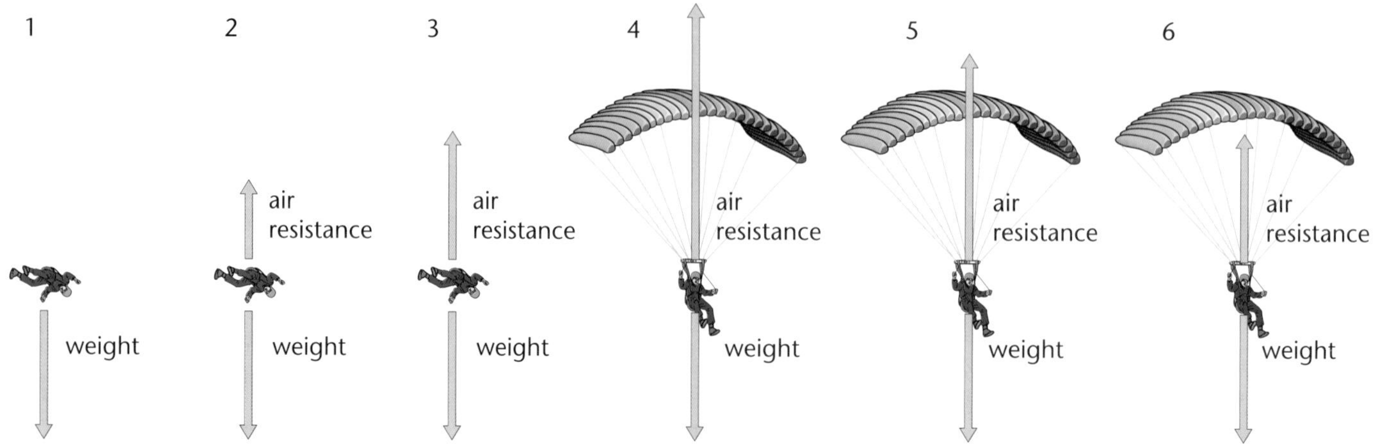

The diagram at the bottom of page 106 show the forces for stages 1–6. The table gives the size of the weight and the air resistance on the skydiver at the same stages during her fall.

b **Copy and compete the table showing the resultant forces and movement of the skydiver. The first row is done for you.**

Putting on the brakes

Normal cars do not have parachutes to show them down! They have brakes. Brakes increase friction. Formula 1 racing cars have to slow down very quickly. The photo shows the car brakes glowing. If you rub your hands together, your palms get hot. The kinetic energy ends up as thermal energy. The same thing is happing in the car's brakes, only there is a lot more kinetic energy to be transferred.

Ships don't have brakes. They rely on the drag between the water and the ship to slow them down. This works better than it would with a car, because there is more friction between water and the sides of a ship than between air and the sides of a car. Big ships like oil tankers need a huge stopping distance – 20 miles or so!

The space shuttle also has no brakes. It uses the friction between itself and the atmosphere to slow it down on re-entry. Again, like the racing car, this generates a lot of thermal energy, raising the temperature of the surface to 3000 °C. The heat shield on the outside of the shuttle insulates the astronauts.

Did you know?

You can investigate how air resistance changes with speed now! Push the flat of your hand through the air. If you push slowly, you feel nothing. If you push fast, you feel the air that you are pushing out of the way.

Questions

These questions refer to this speed–time graph, which shows a parachutist falling.

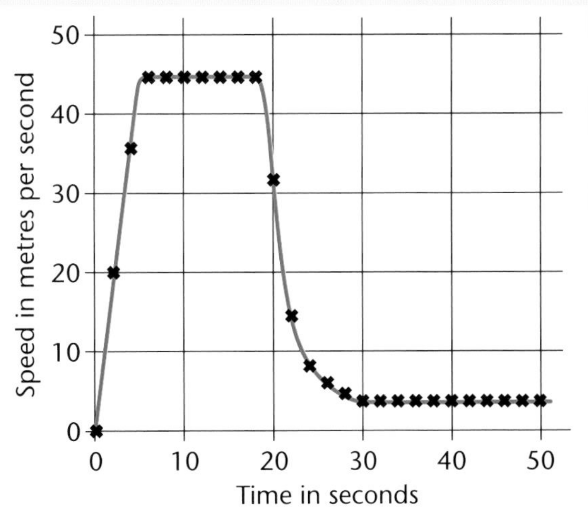

1 What is:

a the force upwards **b** the force downwards on the parachutist?

2 Between which times during the fall are these two forces equal? Explain how you made your decision.

3 Between which times during the fall is the parachutist:

a accelerating? **b** decelerating?

4 Explain why the parachutist has a larger air resistance with his parachute open than closed. Use the word 'particles' in your answer.

5 Explain why the air resistance is greatest just after the parachute opens, and then decreases as the parachutist slows down.

c **A Formula 1 racing car has brake pads made of carbon. These brake pads 'vanish' during the race. Explain why.**

For your notes:

- Friction slows things down. Increasing friction can stop moving objects.
- Brakes use increased friction and parachutes use increased air resistance to slow things down.
- Parachutes have a large surface area for the air particles to hit.

L1 Under pressure

Sinking feeling

Karl and Jackie both have the same weight. They are pushing on the snow with the same force. Karl's boots sink into the snow while Jackie's feet stay on the surface and do not sink.

a Why does Karl sink into the snow?

Jackie is wearing snowshoes so her weight has been spread out over a larger area. We say that the **pressure** beneath Jackie's feet is lower than the pressure beneath Karl's feet. Karl **exerts** more pressure.

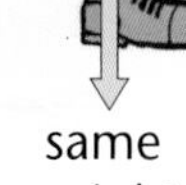

Pressure at the sharp end

When you push a drawing pin into a board, the force from your thumb on the drawing pin is concentrated at its sharp end. The pressure at the sharp end is very high because the area is so small. The pin is able to move into the board.

Look at this photo. The feet of a camel are quite large and flat. The contact area between the camel and the sand is large. This means that the pressure is quite low, so the camel does not sink into the sand.

b Why does a camel find it easy to walk on sand?

The area of contact between a sharp knife and a piece of cheese is very small. This means that the pressure is very high. The knife is able to cut into the cheese quite easily.

c Explain why a knife does not cut very well when it is blunt.

What is pressure?

You can see from the examples above that pressure depends on the force acting and the area that it is acting on. For the same force:

- as the area increases, then the pressure decreases
- as the area decreases, then the pressure increases.

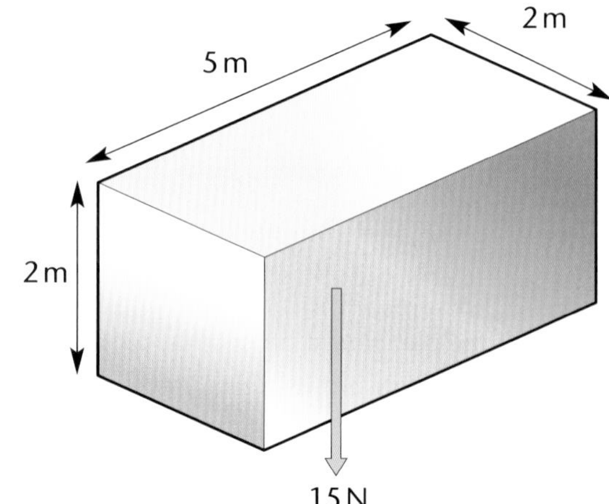

The rectangular block in this diagram has a weight of 15 N. It is lying on its side with its largest face on the floor. The area of this face is $10\,m^2$ ($2\,m \times 5\,m$).

You can calculate the pressure under the face of the block by using this equation:

$$\text{pressure (in N/m}^2) = \frac{\text{force (in N)}}{\text{area (in m}^2)}$$

For the rectangular block on the previous page:

$$\text{pressure} = \frac{15\,\text{N}}{10\,\text{m}^2} = 1.5\,\text{N/m}^2$$

Pressure is measured in units of **newtons per square metre** (N/m^2). This unit is also called a **pascal** (Pa).

Now the block is placed on its end. The weight is still 15 N, but the surface area in contact with the floor is smaller.

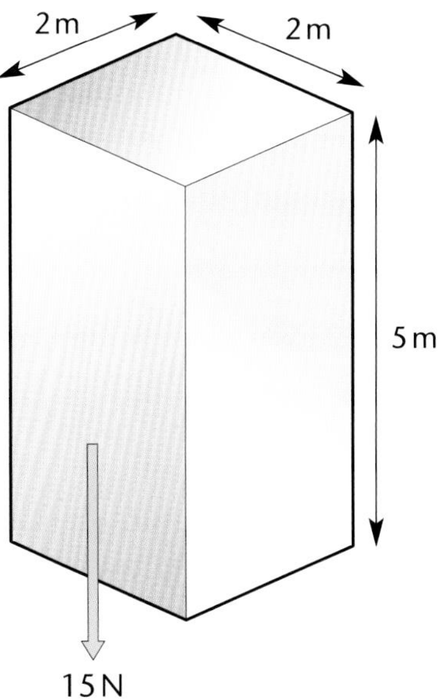

d Calculate the pressure under the block in this diagram. What has happened to the pressure now?

e A woman standing exerts a force of 600 N over the area of her two feet, 0.005 m^2. An elephant standing exerts a force of 60 000 N over the area of its four feet, 0.8 m^2. Which of them exerts the higher pressure?

Changing things around

Sometimes you may want to calculate the force or the area rather than the pressure. You can use the pressure triangle to help you with pressure calculations.

You put your finger over the thing you want to calculate, say force, and the rest of the triangle shows pressure multiplied by area. So to find the force you multiply the pressure by the area. In the same way, to find the area, you divide the force by the pressure.

f If the pressure under a block is 4 N/m^2 and the area of the bottom of the block is 0.4 m^2, what is the force exerted by the block?

Questions

1 Copy the table on the right and use the pressure triangle to complete it.

	Force	Area	Pressure
a	30 N	10 m^2	
b	40 N	0.5 m^2	
c	10 N	2 m^2	
d		4 m^2	2 N/m^2
e	20 N		0.5 N/m^2

2 Jackie was watching some people trying to rescue a child who had fallen through the ice into a lake. She noticed that one of the rescuers lay flat on the ice and moved forward slowly. Explain why he did this instead of walking across the ice.

3 Use your knowledge of pressure to explain:

- **a** why you will bleed if you push down on the pointed end of a drawing pin, but not if you push down on the flat end
- **b** why the wheels of a tractor need to be large and wide but the blades on the plough it is pulling are very thin
- **c** how a woman can damage a wooden floor when wearing stiletto heels but not when she wears trainers.

4 The area of one of Daniel's shoes is 0.02 m^2. His weight is 800 N.

- **a** What is the pressure beneath Daniel if he stands on one leg?
- **b** What is the pressure beneath Daniel if he stands normally?
- **c** Daniel stands on his toes. Does the pressure increase or decrease? Explain your answer.

For your notes:

- The **pressure** depends on the force of an object and the area over which the force is acting.
- Pressure is calculated using the equation:

$$\text{pressure (in N/m}^2) = \frac{\text{force (in N)}}{\text{area (in m}^2)}$$

- For a given force, as the area increases the pressure **exerted** decreases, and as the area decreases the pressure exerted increases.

L2 Taking the plunge

Learn about:

- Pressure in liquids

Earth movers

Machines like the digger in this photo can move or lift heavy things very easily. The digger has two big syringes that move and put pressure on the liquid inside them.

The big squeeze

Look at the diagram below showing two syringes filled with water. They are joined together by a plastic tube. When plunger A is pushed in, the liquid is put under pressure. Plunger B is pushed out. The pressure has been **transmitted** through the liquid from plunger A to plunger B. The pressure is the same throughout the liquid, because it acts equally in all directions. The force on plunger B is the same as the force on plunger A.

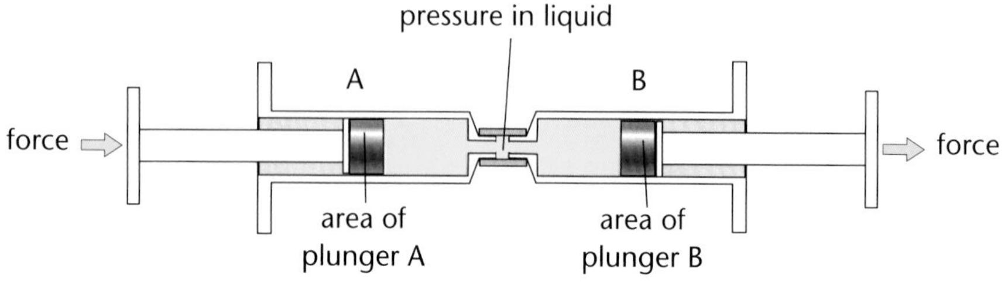

a **Explain what happens in a liquid when you try to squash it. Use the particle model to help you.**

Different sizes

Think what would happen if the two plungers were different sizes. The pressure transmitted is still the same, but the force on each plunger is different. Look at this example on the right. Plunger A is pushed in with a force of 10 N. Because the pressure on plunger B is the same, we can work out the force that pushes out plunger B.

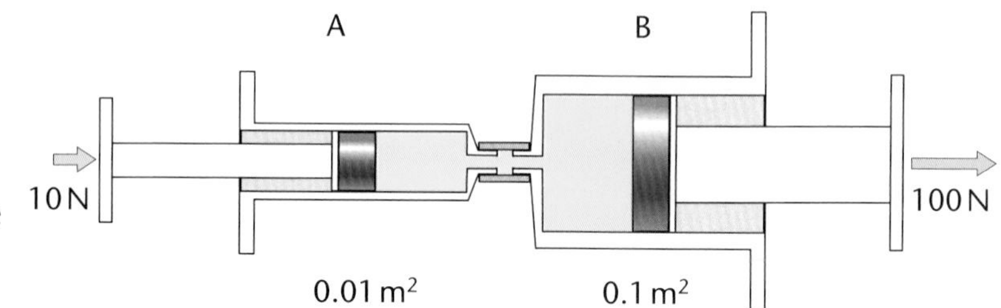

$$\text{pressure on A} = \frac{\text{force}}{\text{area}} = \frac{10}{0.01} = 1000\,\text{N/m}^2$$

$$\text{force on B} = \text{pressure} \times \text{area} = 1000 \times 0.1 = 100\,\text{N}$$

b **Compare the force on each plunger, and their areas.**

If you exert a force of 10 N on plunger A, then plunger B has a force of 100 N on it. The force has increased. A hydraulic system **magnifies** the force you put into it.

$$\frac{\text{small force}}{\text{small area}} = \frac{\text{large force}}{\text{large area}}$$

Hydraulics

The two syringes make up a machine called a **hydraulic machine**. A hydraulic machine magnifies the force to make it easier to lift heavy things. The liquid-filled syringes are called **cylinders** and the moving plungers are **pistons**. If a small force acts on a small input piston, a much larger force acts on a larger output piston connected to it. Many machines, including diggers, use hydraulics.

Did you know?

'Hydraulic' comes from the Greek word for water, as many hydraulic systems use water, though some use oil.

If the area of the output piston is twice the area of the input piston, then the output force will be twice as big as the input force. If the area of the output piston is 10 times bigger, then the output force will be 10 times bigger. This is shown in the table below.

Input piston		Pressure in system in N/m²	Output piston	
Force in N	Area in m²		Force in N	Area in m²
100	0.1	1000	1000	1

You only need a force of 100 N to lift up a heavy weight of 1000 N.

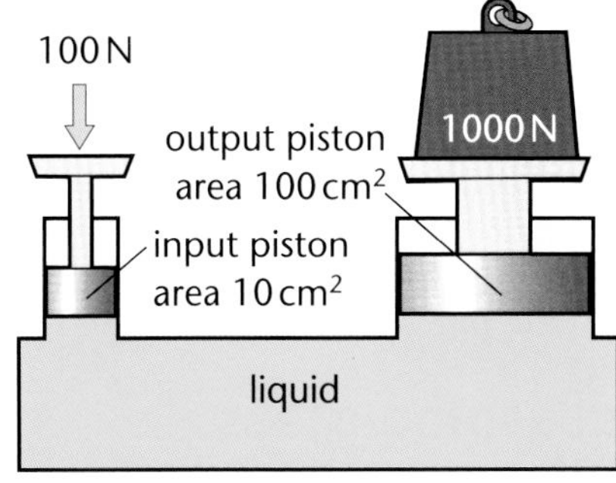

You calculate the pressure on the input piston. Then you use the pressure triangle to calculate the force on the output piston, or its area.

Water pressure

So far, we have thought about putting pressure on a liquid from outside. But a liquid also has its own internal pressure. You may have experienced this if you have swum deep down in a swimming pool. Sometimes this pressure makes your ears hurt. The deeper you go, the heavier the weight of water there is over you, pushing down on you. This force over the area of your body exerts a pressure called **water pressure**.

Did you know?

People can't dive deeper than about 120 metres in water because the water pressure would crush them. When you get to 10 000 metres under the ocean, the pressure is equivalent to seven elephants standing on one plate!

Questions

1 Copy and complete this table. Each row shows a different hydraulic machine.

Input piston		Pressure in system in N/m²	Output piston	
Force in N	Area in m²		Force in N	Area in m²
10	2	**a**	**b**	4
10	0.1	**c**	50	**d**
e	0.2	**f**	300	2

2 In a digger arm, the area of the output piston is 20 times bigger than the area of the input piston. A force of 25 N acts on the input piston. What is the force on the output piston?

3 A car brake system has one input piston connected to the brake pedal and four output pistons connected to the four wheels. The driver pushes the brake pedal with an input force of 10 N. The area of the input piston is 0.01 m². The area of each output piston is 0.05 m².

a How much force is exerted on one output piston?

b What is the force exerted on one wheel?

c What is the total force exerted on all four wheels?

4 Cheryl the milkmaid weighs 500 N and Ermintrude, her cow, weighs 5000 N. Design a hydraulic system that would allow Cheryl to raise the cow in the air to make milking easier.

5 Explain why the deeper you dive into a pool the more likely you are to damage your ears.

C **Submarines dive much deeper than 120 metres under the ocean. What design features do you think they have to cope with the water pressure?**

For your notes:

- Liquids cannot be squashed. The pressure in a liquid acts equally in all directions so the pressure is **transmitted**. A **hydraulic machine** uses this property of liquids.
- A hydraulic machine **magnifies** the force. If a small force acts on a small input **piston**, a much larger force acts on a larger output piston connected to it.
- Water has its own pressure. The deeper you go, the heavier the weight of water so **water pressure** increases.

L3 Pressure in the air

Learn about:

- Pressure in gases

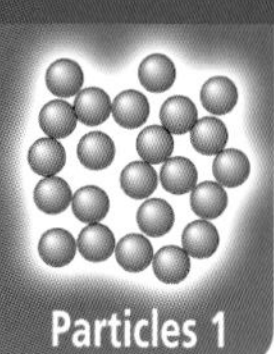

Pump it up!

If you have ever played with an old syringe or a bicycle pump, you know what happens if you put your finger over the hole and then push the plunger in. You exert a force on the plunger and the plunger moves in a little.

a **Use your knowledge of the particle theory to explain what happens to the particles in a gas when the gas is squashed, and why this happens.**

Squashing gases

When you **compress** (squeeze) a gas in a syringe, the volume it takes up decreases. The number of gas particles doesn't change.

- When the volume decreases, the pressure increases.
- When the volume increases, the pressure decreases.

The pressure also increases if you pump more air into a fixed volume, because there are more gas particles in the fixed volume. If there are more particles, they will be closer together and the pressure will be higher.

pressure gauge

N/m²

N/m²

Do you remember?

The particles in a gas are moving around in all directions. They hit the sides of their container and the force caused by all these collisions is gas pressure.

b **Think about what happens when you pump up a bicycle tyre. Compare the pressure in the tyre before and after you pump it up. Explain why there is a difference.**

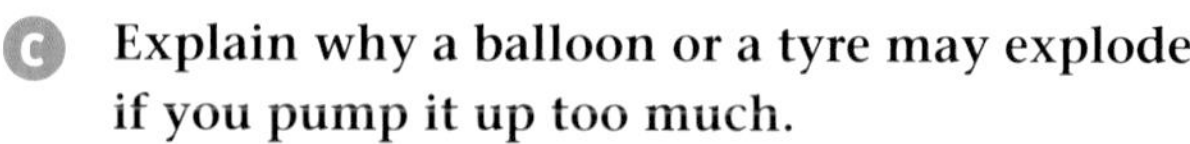

c **Explain why a balloon or a tyre may explode if you pump it up too much.**

We use gas pressure in many ways. Think about a rubber suction pad which sticks to smooth surfaces. When you press the suction pad into place, the rubber flexes and some of the air inside it is squeezed out from underneath. When you let go the rubber springs back, so the volume under the pad increases. The pressure under the pad is now lower than the air pressure outside. This difference in pressure gives enough force to hold the pad in place.

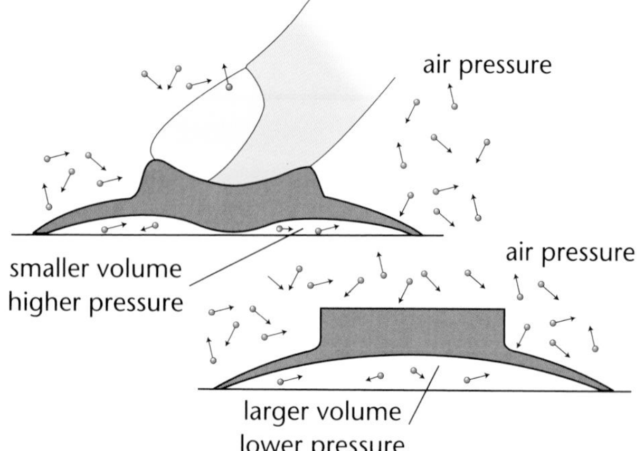

Releasing pressure

If you compress a gas into a small volume and then release it, it expands quickly to fill a larger volume again. It expands until the pressure inside the container is the same as the air pressure outside the container. This expansion of gases is used in aerosols of whipped cream. Inside the can is a mixture of cream and pressurised carbon dioxide. When you push down on the nozzle, the volume gets bigger and the carbon dioxide expands very quickly. A little of the gas, mixed with the cream, rushes out very quickly.

Pneumatics

The sudden expansion of a compressed gas is also used in machines like hydraulic machines, with cylinders and pistons. These are called **pneumatic machines.** A pneumatic drill like the one in this photo has a very fast-moving piston that can move many times a second. It is used for breaking up pavements and drilling holes for explosives. An electrical pump, called a compressor, forces the air at high pressure into a cylinder. The air expands and pushes the piston down. The piston hits the drill bit and forces it down into the pavement.

Did you know?

'Pneumatic' comes from the Greek word for air or breath.

Air pressure

Like water, air has internal pressure which is called **air pressure**. The air pressure changes with depth, just as water pressure does. The air pressure is greatest on the ground at sea level, because there is a large weight of air overhead pushing down. The higher up in the air you go, the lower the pressure becomes, because the weight of air decreases. The air pressure in high mountains is half the air pressure at sea level.

Aeroplanes fly at about 15 000 m above the ground where the air pressure is even lower. You cannot breathe at this height because there is too little air. The inside of an aeroplane is pressurised to make the pressure approximately the same as it is on the ground, so that the passengers can breathe.

d **Imagine you take an empty plastic bottle with you in an aeroplane when you take off. When the plane reaches its cruising height the sides of the plastic bottle have caved in a little. What does this tell you about the air pressure in the plane in compared with that at the airport?**

e **The air pressure exerted by the atmosphere is about 100 000 N/m^2 at ground level. If you assume that the area of your head is 0.1 m^2, calculate the force on your head.**

Questions

1. When a balloon has a slow leak, the number of air particles inside is decreasing and they are becoming more spread out. Is the pressure inside the balloon increasing or decreasing?
2. Explain the following statements.
   - **a** A rubber suction pad can be used to unblock sinks.
   - **b** Air pressure gets lower the higher up a mountain you go.
   - **c** Aeroplanes have pressurised cabins.
3. In a carefully controlled experiment, a can with air inside it is sealed and heated.
   - **a** Describe how the particles move inside the can before heating.
   - **b** Explain what happens to the particles inside the can after heating.
   - **c** What do you think might happen if the can is heated strongly for a long time?
4. Suggest a reason why a mountain climber has difficulty breathing near the top of the mountain.
5. Your school's technology department is putting on an exhibition about pneumatics for a Year 6 class from the primary school. Choose one pneumatic machine and design a leaflet explaining how it works for them.

For your notes:

- Gases can be **compressed** because there is space between the particles. This is called gas pressure.
- When a gas is compressed, the volume decreases and the pressure increases. When a compressed gas is released, the volume increases and the pressure decreases.
- The fast expansion of released gases is used in **pneumatic machines** and aerosols.
- Air has internal pressure. The higher up in the air you go, the lower the **air pressure** becomes, because the weight of air decreases.

L4 Where's the pivot?

Learn about:

- Pivots and levers

Shut the door

Imagine that you are closing a door. First you try to push it very close to the hinge, as shown in the picture on the right. Now imagine you push the door shut at the handle. This is much easier. The further away from the hinge you are, the less force you need to turn the door around the hinge.

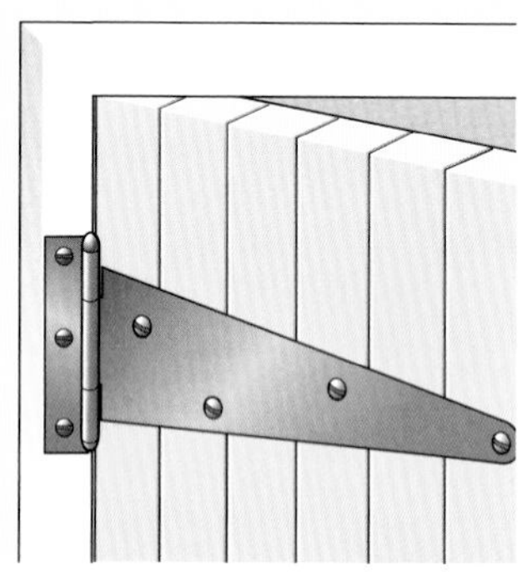
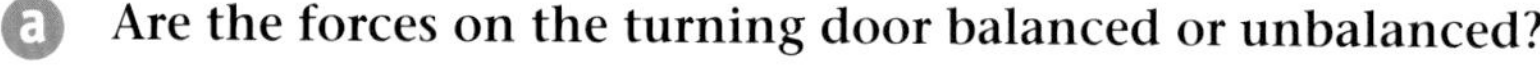

Pivots are everywhere

Forces can make things turn. A door turns around the hinge. A wheel turns around the axle. The **pivot** is the point around which something turns. When you open the door the hinge turns very little but the whole door turns more.

Hinges are pivots.

Do you remember?

Unbalanced forces can make things move forwards or backwards or stop moving.

a **Are the forces on the turning door balanced or unbalanced?**

The joints in your skeleton also act as pivots.

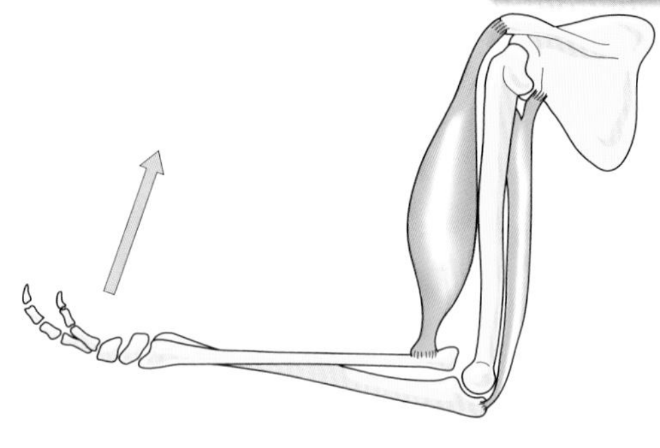

People use machines to do jobs more easily. Some machines help you lift things using less force. The force of the object you are lifting is called the load. The load is the object's weight.

A wheelbarrow is a machine. It also uses a pivot. When you lift a weight in a wheelbarrow, the wheel acts as a pivot. The wheelbarrow helps you lift up the weight more easily than if you lifted it straight up without the wheelbarrow.

b **Draw a simple diagram of the person using the wheelbarrow. Label the pivot, the load and the person's force acting on the wheelbarrow.**

If you used to sit on a seesaw with a friend on the other end, you'll remember that you can lift the friend more easily if you sit far away from the pivot than if you sit close to it.

Holly and Cameron both have the same weight. If they both sit at the ends of the seesaw, they are balanced. But if Holly moves towards the pivot, her end of the seesaw moves upwards. They are unbalanced.

c **Why does Holly's end move upwards when she is closer to the pivot?**

d **If Cameron were heavier than Holly, where should he sit to balance her when she sits at the end?**

Levers

To move something around a pivot, you use a **lever**. When you close or open a door, the door is a lever. With a wheelbarrow, the handles and body of the wheelbarrow are the lever. As you can see with the door and the seesaw, the further away from the pivot the force acts, or the longer the lever, the less force you needed to turn the lever.

e **Look at these photos of opening a tin and using a crowbar. For each one, say what the lever is and where the pivot is.**

Round in circles?

If you think about a lever moving, you will see that the lever is turning around the pivot. The force on the end of the lever is having a **turning effect** which turns the lever through part of a circle. It may be difficult to see because some levers only turn through a small part of the circle. The force at the end of the lever acts in a straight line.

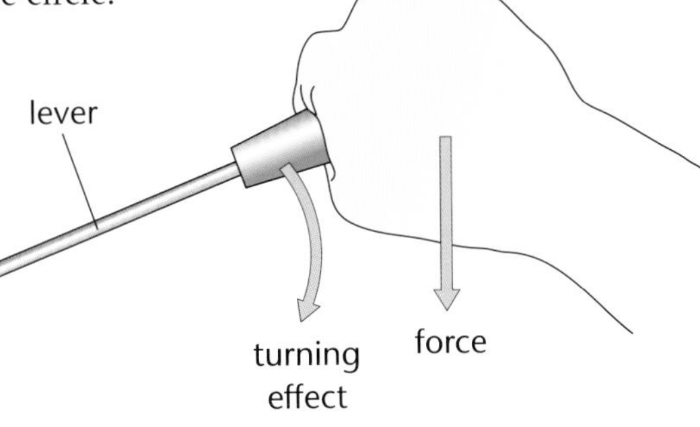

The turning effect of a force depends on the on the size of the force. It also depends on the distance between the pivot and the force. If the force is acting directly towards the pivot then nothing will turn. This is shown on this diagram of the tap – the forces marked ✗ have no turning effect on the tap.

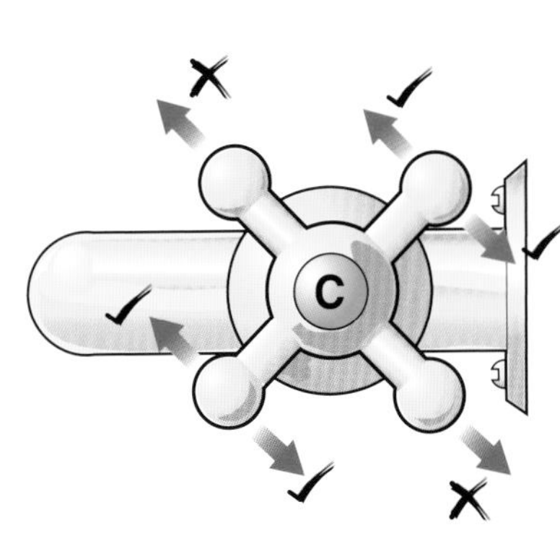

Questions

1 Look at the photo of a crowbar above. Draw a simple diagram to show the lever, the pivot, the force acting and the turning effect.

2 Sam is having difficulty opening a paint tin with a screwdriver. He decides to use a longer screwdriver. Explain why.

3 If two people are sitting on a seesaw, each side acts as a lever. The seesaw is a fixed length. How can you make the levers act as if they are shorter or longer?

4 The turning effect of a lever depends on two variables. What are they?

5 People with arthritis in their hands find it hard to grip things and apply enough force to screw and unscrew things. What could a designer do to help them with this problem?

6 Make a worksheet to explain to primary school children why levers are useful machines. Include a simple activity for the pupils to carry out and provide answers for any questions you set.

For your notes:

- A **pivot** is the point around which an object such as a door or a crowbar turns.
- When a force acts on a **lever** it makes the lever turn around the pivot.
- The **turning effect** of a force depends on the size of the force and the distance between the pivot and the force.

L5 Balancing act

Learn about:
- Moments

Rotating doors

Have you ever tried to push a rotating door with someone else pushing harder against you on the other side of the door? You are pushed backwards and can't get in.

To turn a lever around a pivot, the forces on the lever must be unbalanced. Cameron is pushing the rotating door with a force of 50 N. Mahir is pushing it harder in the opposite direction, with a force of 60 N, so the door is turning. The turning effect of their forces are unbalanced.

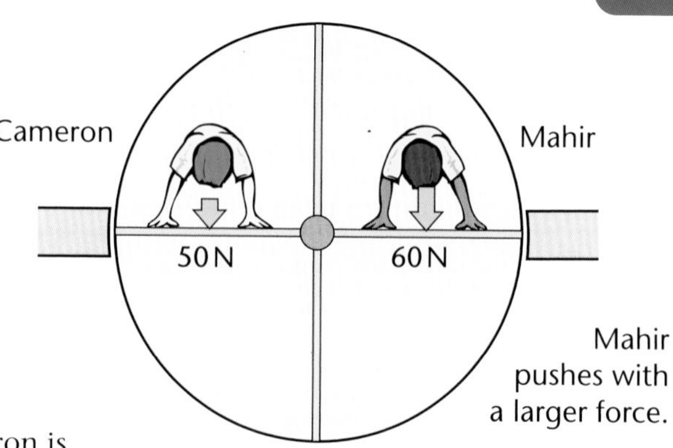

Mahir pushes with a larger force.

Sometimes the turning forces may be balanced. In this picture Cameron is pushing the rotating door in one direction with a force of 50 N. Holly is pushing against him in the opposite direction also with a force of 50 N. The forces exerted by Cameron and Holly are balanced. Their two forces have a balanced turning effect on the door. The door does not turn!

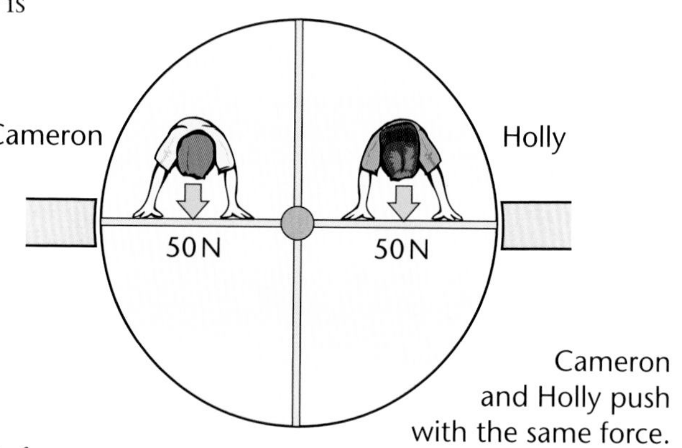

Cameron and Holly push with the same force.

a **Look at these two diagrams of people pushing on the rotating doors. What can you say about the distance from the pivot that they are pushing?**

Moments

Cameron's force acts on the door in an anticlockwise direction. Holly's force acts on the door in a clockwise direction. There are two forces. They are the same size and act in opposite directions, so they are balanced.

The turning effect of a force around the pivot of the door depends on both the force acting and the distance from the pivot to the force. The distance from the pivot is measured perpendicular to the direction of the force.

The turning effect of a force is also called the **moment** of the force. We measure force in newtons, and distance in metres, so the units for a moment are newton metres, Nm. This is summarised in this equation:

moment of a force	=	force	×	distance
in Nm		in N		in m

Do you remember?

When the forces on an object are balanced, the object will stay still or it will move at a steady speed.

distance perpendicular to force

1.5 m

50 N

In the diagram of the revolving door above, Cameron is exerting a force of 50 N. The force is 1.5 m from the pivot.

moment of Cameron's force = force × distance
= 50 N × 1.5 m
= 75 Nm

b **What is the size of the moment around the pivot when Holly pushes the door?**

When Cameron and Holly push the door, the two moments are balanced. Cameron's anticlockwise moment is equal to Holly's clockwise moment. The **principle of moments** says that when something is balanced:

the sum of the anticlockwise moments = the sum of clockwise moments

Balanced moments

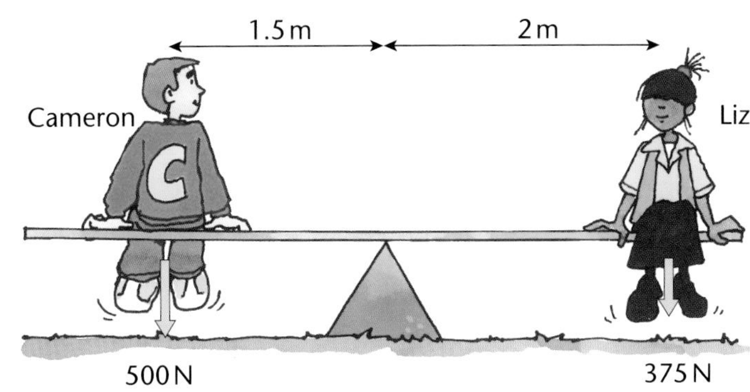

On this seesaw Cameron's weight is 500 N and Liz's weight is 375 N. Cameron is 1.5 m from the pivot and Liz is 2 m from the pivot. We can put these numbers into the equation to find the moments on each side. Remember that each moment is calculated as force × distance.

anticlockwise moment = clockwise moment
force × distance for Cameron = force × distance for Liz
500 N × 1.5 m = 375 N × 2 m
750 Nm = 750 Nm

As you can see the moments are the same, so they balance.

Unbalanced moments

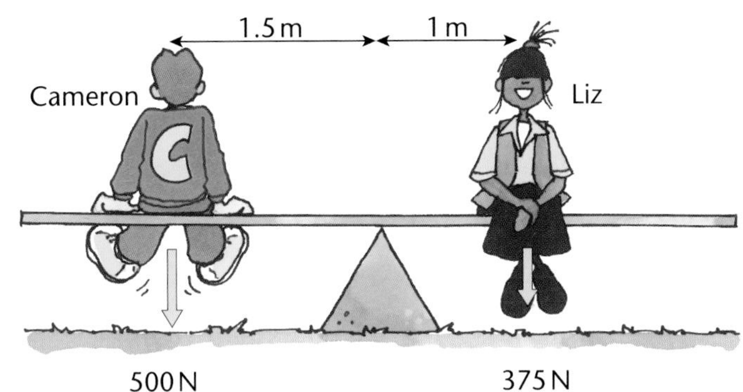

Now Liz moves towards the pivot of the seesaw, and it tips. We can do the same calculations to see what the moments are now and why they are unbalanced.

anticlockwise moment = clockwise moment
force × distance = force × distance
500 N × 1.5 m   375 N × 1 m
750 Nm   375 Nm

As you can see the moments are not the same, so they don't balance.

c Which is greater, the clockwise moment or the anticlockwise moment?

d Which way will the seesaw tip?

Questions

1 Mike sits at one end of a seesaw and balances Ken at the other end. Both of them weigh the same.

a What can you say about the distance the two boys are from the pivot?

b What would happen if one of the boys was heavier?

2 Explain the following:

a the moment of a force

b what happens when the anticlockwise moment is bigger than the clockwise moment.

3 Copy the table and complete it by calculating the missing values.

Example	Force in N	Distance	Moment of force in Nm
A	10	1 m	?
B	?	50 cm	25
C	55	?	38.5

4 a Look at this diagram. Calculate the anticlockwise and clockwise moments.

b Will the seesaw balance or not?

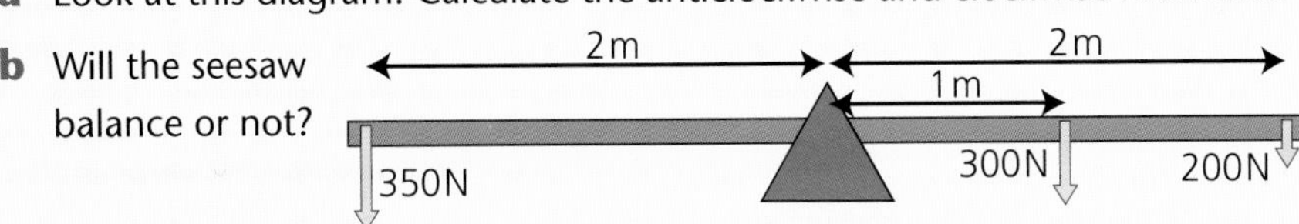

For your notes:

- The turning effect of a force is called the **moment** of the force.

moment of a force in Nm = force in N × distance in m

- When two moments are balanced, an object will not turn and the **principle of moments** applies. This is summarised by the equation:

the sum of the anticlockwise moments = the sum of the clockwise moments

L6 Moments in life

Learn about:

- Using moments

Work it out

You have seen that a lever turning round a pivot is a simple machine that can be very useful in everyday life. We can use the principle of moments to understand how and why everyday machines such as levers work.

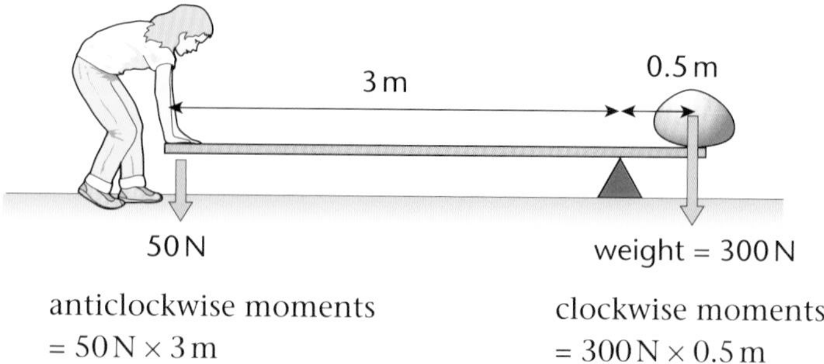

anticlockwise moments
= 50 N × 3 m
= 150 Nm

clockwise moments
= 300 N × 0.5 m
= 150 Nm

If we calculate the moments about a lever, we can see what a difference the lever makes to lifting things. A lever is like a seesaw: there are clockwise and anticlockwise moments. You can see these in the diagram on the right.

You might think that you are getting something for nothing, because you use less force to lift the load than the load itself. But this isn't the case. The load moves through a very short distance when you lift it on the end of the lever, while you have to move the end you are pushing on through a much greater distance. So although you need to use less force, you have to move the lever through a greater distance than the load moves.

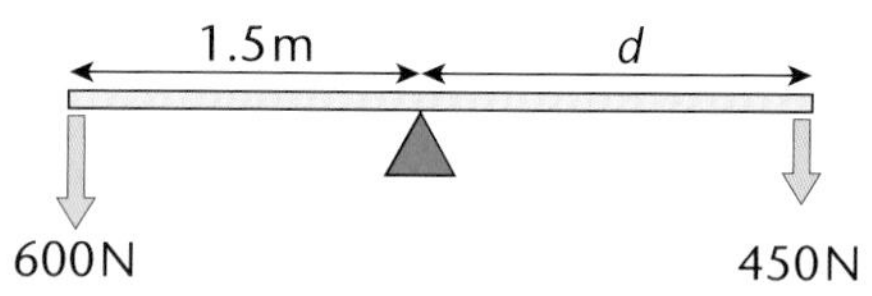

the sum of the anticlockwise moments = the sum of the clockwise moments

$$600\,N \times 1.5\,m = 450\,N \times d$$
$$900\,Nm = 450 \times d$$
$$d = 900/450$$
$$d = 2\,m$$

More calculations

When something is balanced, you know that the clockwise moments equal the anticlockwise moments. So if you don't know one of the values that make up the moments, you can calculate it as long as you know the other values. For example, if you know that a rock weighs 600 N, the rock is 1.5 m from the pivot and you can push it with a force of 450 N, you can calculate how long a lever you would need to hold the rock up.

a **Andy and Caroline are on a seesaw. Andy has a weight of 600 N and is 1 m from the pivot. Caroline's weight is 400 N. How far is Caroline from the pivot if they are balanced?**

b **David sits 2 m from the pivot of a seesaw and balances Liz. Liz weighs 400 N and sits 3 m from the pivot. Calculate David's weight.**

Counterbalance

We have seen how moments working in opposite directions can be either balanced or unbalanced. The principle of moments is important to people who use cranes in building sites or when loading ships. The photo far left shows a crane picking up a heavy load. The crane has a large weight called a **counterbalance** at the other end of the arm. This balances the load and stops the crane falling over.

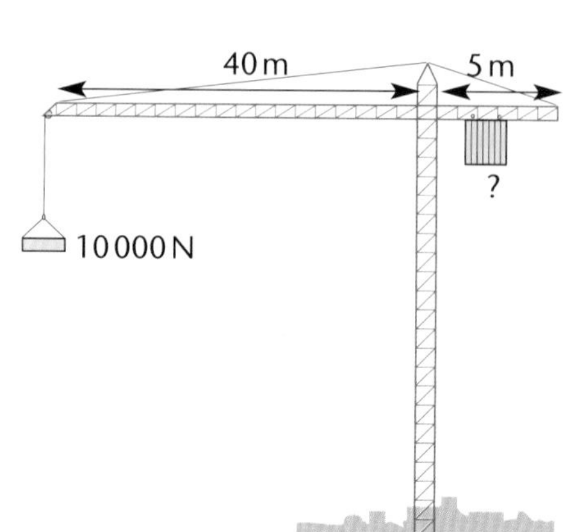

c **Explain why a crane can be used to pick up heavy weights.**

d **Calculate the minimum counterbalance weight needed for the crane shown in the diagram on the left.**

Human pivots

In your body, your bones meet at joints. The joints are the pivots for the bones to turn around. The bones are the levers. If you think about moments you can see that whether you can lift something depends not just on the strength of your muscles but also on how long your bones are, for example, the distance from your elbow to your wrist.

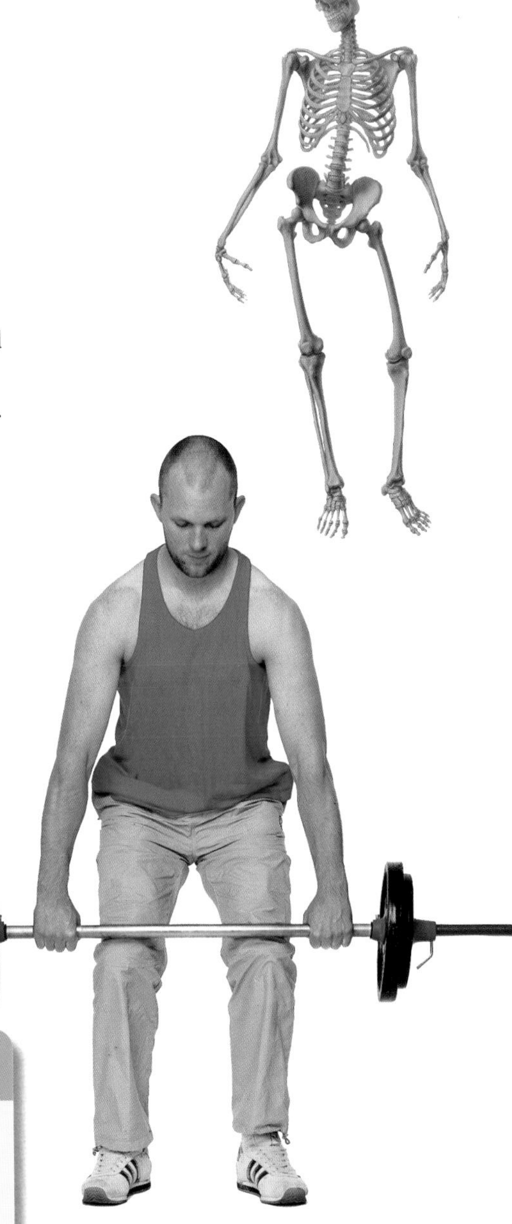

Lifting safely

It is very important to lift and move objects safely. To reach down to the object you want to lift, you must bend your knees. Then use your knees and hips as pivots to help you straighten up, as shown in the photos below. Don't just bend your back to pick up the object. Lifting while bending the back puts a big force directly on the back and can damage it, causing problems such as a slipped disc.

You can make sure you put less strain on your muscles when you carry a heavy object by holding it close to your body.

Questions

1 Explain why holding a weight at arm's length would require more strength than holding it close to your body.

2 Write a paragraph about how to lift objects safely using the words below.

bend **pivots** **knees** **back** **damage**

3 Explain how weighing things with an old-fashioned pair of scales uses the principle of moments.

4 Russ and Kerryanne were on a seesaw which was balanced. Russ weighs 500 N and is 2 m from the pivot. Calculate the distance Kerryanne was from the pivot if she weighs 400 N.

5 **a** Look at the skeleton and list as many places as you can where there are pivots.

b Which bones in the body make the best levers?

6 Design a leaflet for teachers in school about how to lift things safely. Make sure you explain the reasons for lifting in a particular way. Include some diagrams.

For your notes:

- The principle of moments can be used to explain many everyday situations. You can use it to calculate how to make something balance.
- A **counterbalance** is a weight which stops something falling over.

L7 Getting balanced

Think about:

- Equilibrium

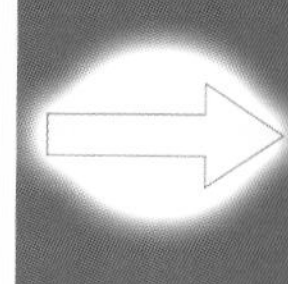

Number balances

Mustapha was using a beam balance. He wondered why it balanced sometimes but not at other times. He experimented with 1 N weights. When the beam is balanced, it is in **equilibrium**.

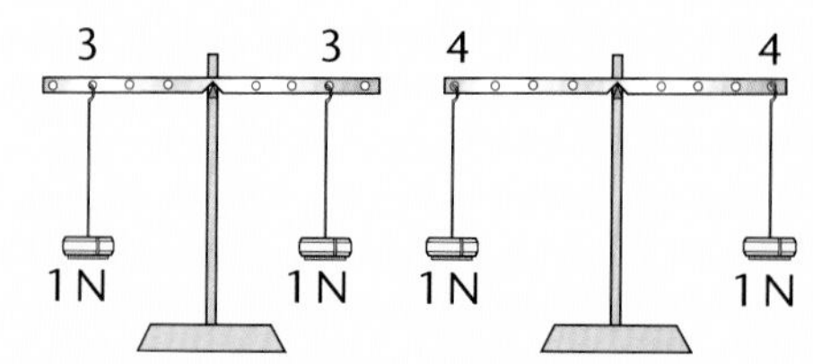

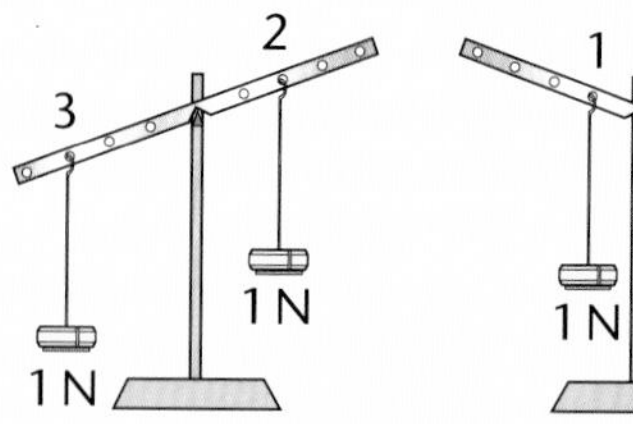

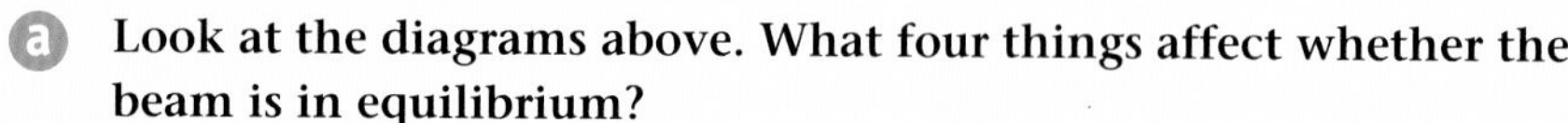

a **Look at the diagrams above. What four things affect whether the beam is in equilibrium?**

Next Mustapha decided to use 4 N and 8 N weights and see what happened. He made this table to show what combinations of weights he used.

The beam balance is working just like a seesaw. We can calculate moments for the left-hand side and right-hand side just as we did for the anticlockwise and clockwise moments of a seesaw. The weight is the force. The numbered holes in the beam give the distance.

	Left-hand side		Right-hand side	
	Distance	**Weight in N**	**Distance**	**Weight in N**
A	1	8	2	4
B	2	8	4	4
C	2	4	1	8
D	4	4	2	8

b **Calculate the moment for each side of the balance in the table and check whether the beam will balance for each of rows A to D.**

c **Describe the relationship between the distance and weight on one side and the distance and weight on the other side in these four examples.**

A bigger balance

Next Mustapha did an investigation using a bigger balance with eight holes on either side. He decided to investigate what happened when he hung weights at two or three different holes.

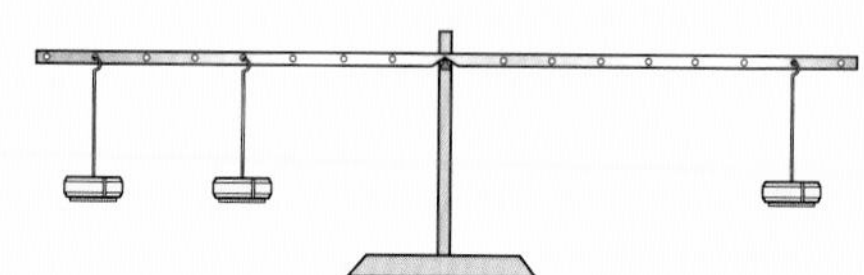

d **Mustapha's second table is shown below. It shows the weights and distances when the beam is balanced. Copy and complete the table, filling gaps i to viii.**

Left-hand side		Right-hand side		
1 N in hole 5		2 N in hole 2	i	
1 N in hole 8		1 N in hole 4	ii	
3 N in hole 7		iii	iv	
2 N in hole 5	3 N in hole 2	4 N in hole 3	v	
1 N in hole 7	4 N in hole 5	5 N in hole 5	vi	
6 N in hole 8	2 N in hole 4	8 N in hole 7	vii	
3 N in hole 3		2 N in hole 1	2 N in hole 2	viii

Equilibrium

As you have discovered, the combination of distance and weight on the left-hand side of the beam has to be equal to the combination of distance and weight on the right-hand side of the beam. This means a moment has two variables which you can change, distance and weight. You can change these for both moments, one each side of the balance. There are other situations in life where you can change a combination of four variables.

Hydraulic equilibrium

Look at the simple hydraulic system in the diagram below. There are two syringes containing a liquid. One plunger is being pushed, so the liquid is under pressure. The system is in equilibrium.

The four variables in the system are:

- force on plunger A
- area of plunger A
- force on plunger B
- area of plunger B.

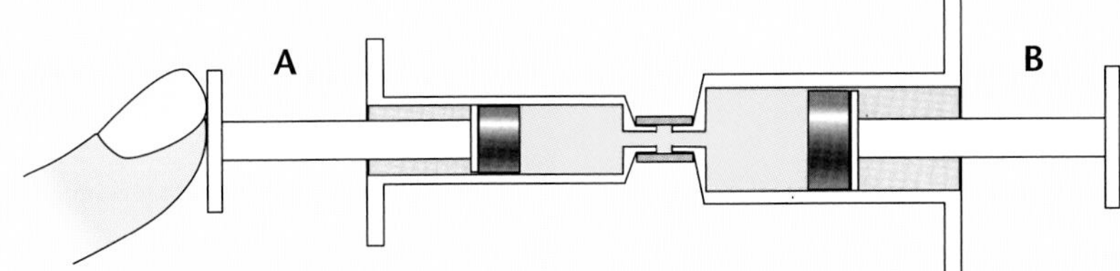

For the system to be equilibrium the pressure caused by plunger A has to be same as the pressure caused by plunger B.

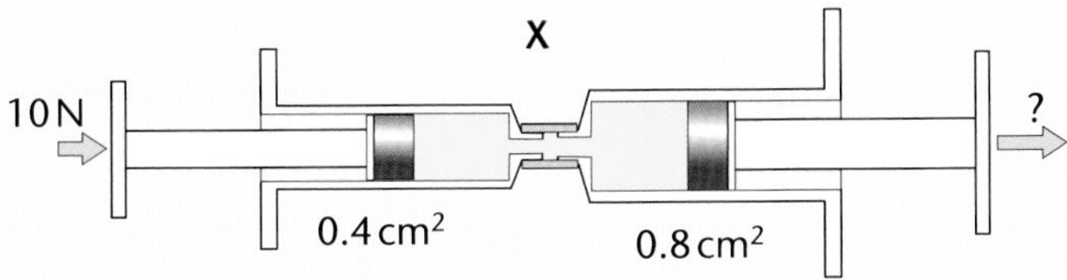

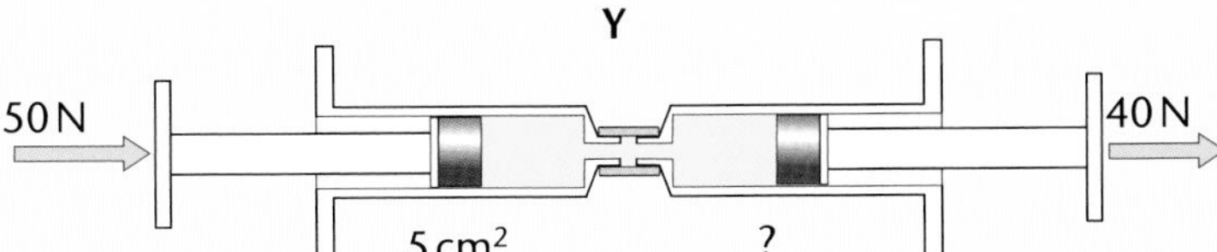

e **Look at the two diagrams above. They each show a hydraulic system in equilibrium.**
(i) In X, predict whether the output force will be bigger or smaller than the input force.
(ii) In Y, is the area of the output plunger bigger or smaller than the area of the input plunger?

f **Calculate the missing values to check your answers.**

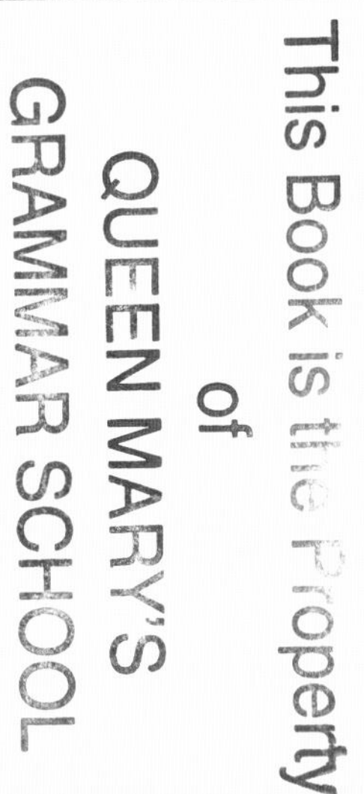

Questions

1 A balance has five holes on either side. There is a 4 N weight in hole 5 on the left-hand side. On the bench there are two hangers, each weighing of 1 N, and 18 1 N weights. Suggest as many ways as possible of making the beam balance without moving the weight on the left-hand side.

2 Explain the meaning of the word 'equilibrium' to your partner, giving examples.

From cells to organs

Focus on:
- Cells

Do you know the basics?

Cells

All living things except for viruses are made up of small building blocks called **cells**. Cells are small – approximately 0.02 mm. There are two main types of cell: **animal cells** and **plant cells**. These have a lot in common, but there are also some differences.

Animal cells

Animal cells, like the one shown below, have three main structures that you have to remember: the **cell membrane**, the **nucleus** and the **cytoplasm**.

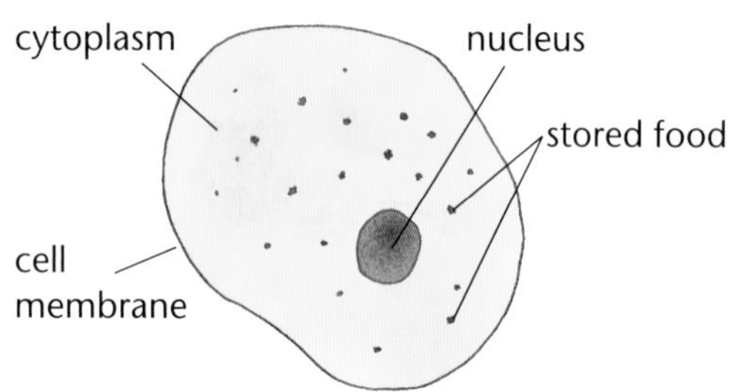

The **cell membrane** lets substances in and out of the cell. The **cytoplasm** is the place where all the chemical reactions take place. The **nucleus** controls everything that happens in the cell.

Plant cells

A plant cell, like the one shown below, also has a cell membrane, a nucleus and cytoplasm. Unlike animal cells, it has a **cell wall**, **chloroplasts** and a **large vacuole**.

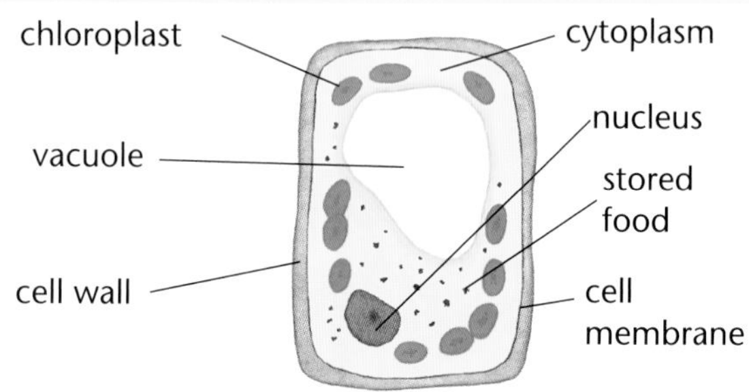

The **cell wall** gives support to the cell and makes it strong. The **chloroplasts** contain chlorophyll. This traps light energy to make food. The cell **vacuole** contains cell sap.

Specialised cells

Many cells are specially adapted. Some examples are: sperm cells, egg cells, root hair cells, epithelial cells, palisade cells and red blood cells.

not to scale

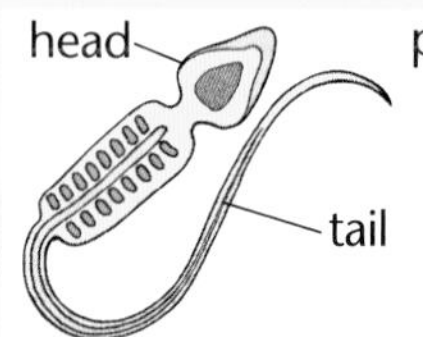

Sperm cells have long tails to swim to the egg and a pointed head which helps them burrow into the egg.

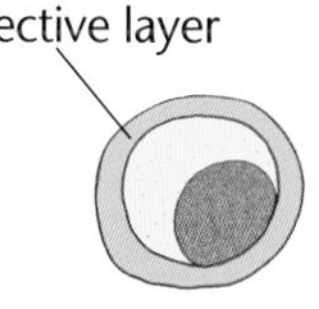

Egg cells have a protective layer so that just one sperm can get through.

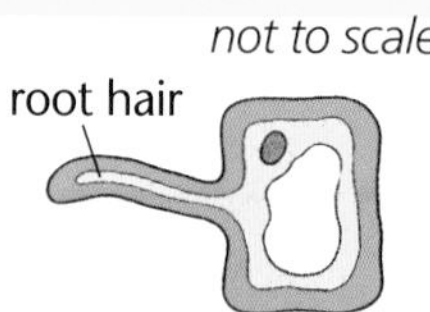

Root hair cells have a long finger-like root hair which gives a very large surface area to absorb water.

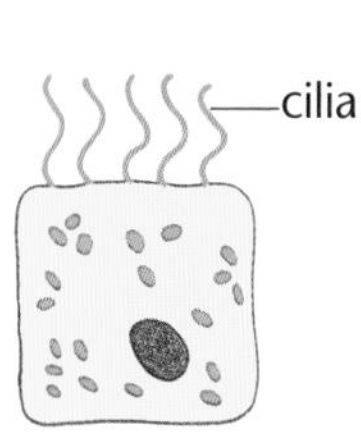

Epithelial cells in nose and throat produce mucus to trap dust and germs.

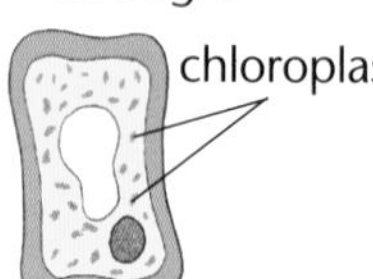

Palisade cells have lots of chloroplasts and are at the tops of leaves to absorb as much light as possible.

Red blood cells have no nucleus and are flexible so they can pass through small blood vessels. They have a large surface area so they can carry lots of oxygen.

Tissues, organs and organ systems

When a group of similar cells carries out a particular function we call the group of cells a **tissue**. When a group of two or more tissues work together they form an **organ**. When a group of organs work together they form an **organ system**.

- Animal tissues include muscle tissue and nerve tissue.
- Plant tissues include onion skin tissue and palisade tissue.
- Animal organs include the heart, lungs, stomach, eyes and brain.
- Plants organs include leaves, stems, roots and petals.
- Animal organ systems include the reproductive system, the digestive system, the respiratory system and the circulatory system.
- A flower is a plant organ system.

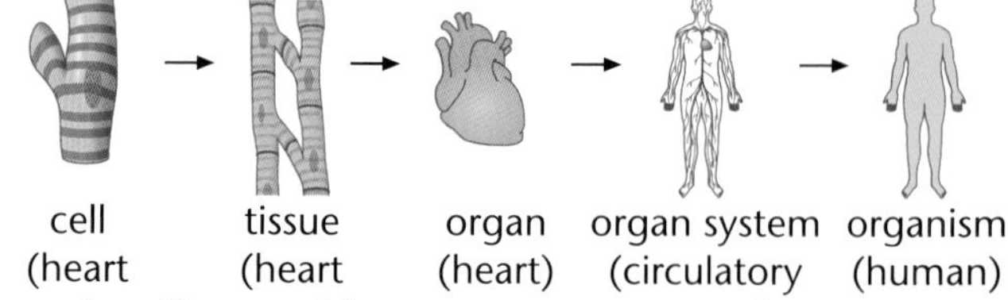

Are you ready for the next step?

a **Explain how you would identify a cell as an animal cell or a plant cell.**

b **Make a table showing two different animal examples of cells, tissues, organs and organ systems, and one plant example of each.**

c **Make a table of the different types of specialised cells and their adaptations.**

d **Which organ systems carry out the following processes?**

(i) reproduction
(ii) the breakdown of food
(iii) the transport of blood around the body

Do you really understand?

All cells **respire** to release energy for life processes.

The success of an organism depends on what happens in every cell, tissue and organ. All the organ systems work together to perform the life processes. There are very many functions for cells to perform, so many cells are specially adapted to particular functions.

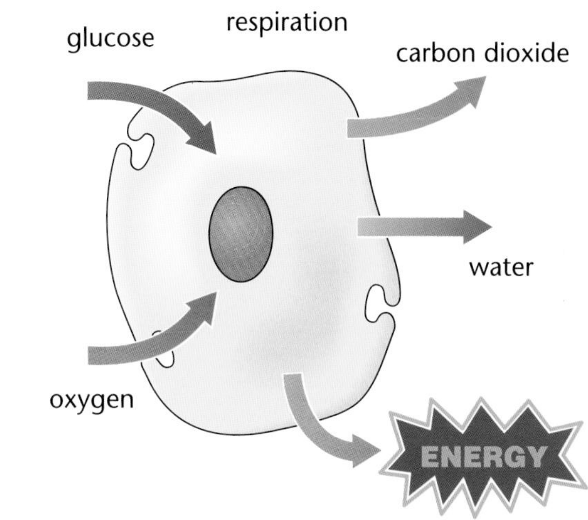

1 Look at the diagram on the right of a single-celled organism called chlamydomonas. It lives in pond water. Its body is one specialised cell.

- **a** Give two features of chlamydomonas that show that it is more like a plant cell than an animal cell.
- **b** Chlamydomonas makes its own food using light. Which part of the cell traps the light?
- **c** Name the parts of the cell that are food stores.

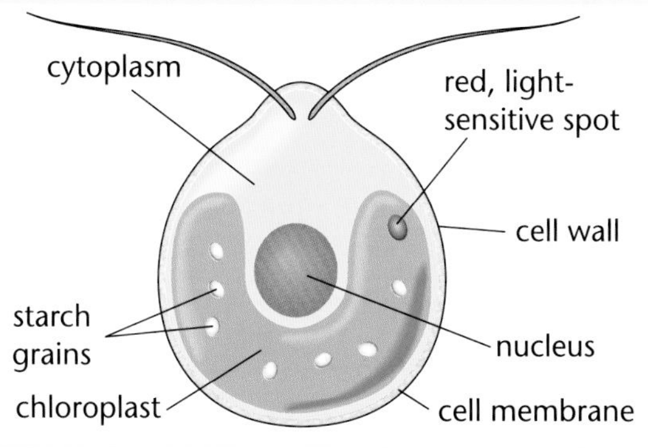

2 Look at the diagram of amoeba on the right. It is another single-celled organism that lives in pond water. Amoeba traps chlamydomonas and digests it.

- **a** Is amoeba a plant or an animal? Explain your answer.
- **b** Amoeba has enzymes to digest starch. What substance is produced from starch and what is it used for?
- **c** In which part of the amoeba are the genes found?

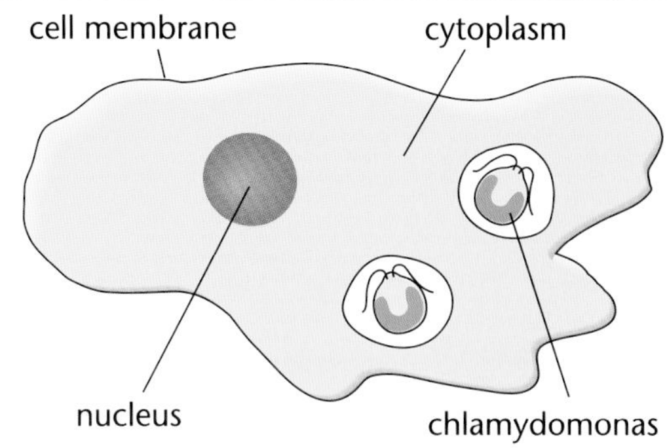

3 Explain what makes the human respiratory system an organ system.

4 Red blood cells pass through small blood vessels to deliver and collect substances for respiration in all cells. Red blood cells are flexible and have a large surface area. Explain how this helps their function.

Interconnections

Focus on:

- Interdependence

Do you know the basics?

Food chains and webs

When animals eat plants, and animals eat other animals, energy flows from one to the other. **Food chains** show the energy transfer between plants and animals. Energy is transferred from a **producer** to a **primary consumer**, then to a **secondary consumer** and so on. Energy enters the food chain when the producer converts light energy from the Sun to chemical energy by photosynthesis.

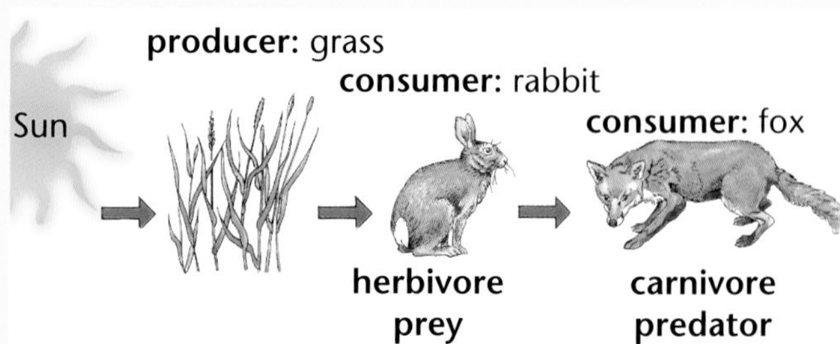

Species in one food chain are often members of other food chains as well. If you draw all the food chains together interlinked in a **food web**, you can see how energy is transferred through all the species that live in a habitat. As the food chains are linked, a change in numbers of one species can affect the numbers of other species in the food web.

Each time energy is transferred from one member of a food chain to the next, energy leaves the food chain. All the biomass at one stage of the food chain does not get converted to biomass in the next stage of the food chain. This is because:

- Some of it is eaten by animals in other food chains.
- Some of it is used for respiration to release energy for movement and other life processes.
- Some of it is lost as heat.
- Some of it is excreted as waste.
- Some of it is not eaten, such as the bones of animals or the hard bits of plants.

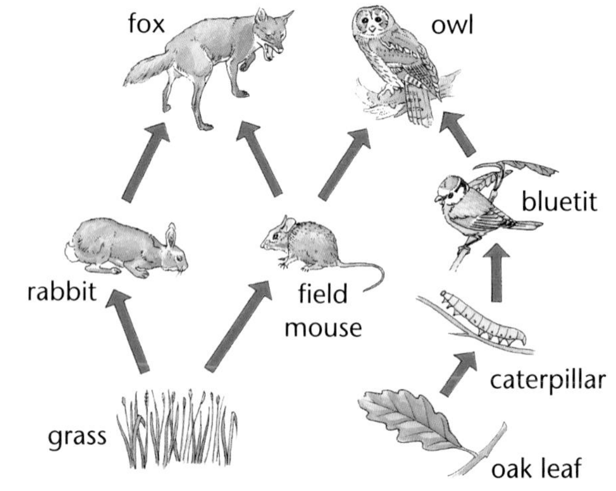

Humans are like any other animal – we are members of lots of different food chains and food webs. Energy enters all food webs through photosynthesis, so humans depend on plants to survive. As in all food chains, a lot of energy has already been wasted before it reaches humans at the end.

Energy is not the only thing that gets passed along the food chain. Some toxins from pesticides are passed along a food chain, and they build up in the top consumers. The total amount of pesticide at each stage of the food chain is the same. But there are fewer organisms towards the end, so the pesticide becomes more concentrated in each one.

Pyramids of numbers

To show the number of each species in a food chain, we draw a **pyramid of numbers**. Bars of different lengths represent the numbers of organisms.

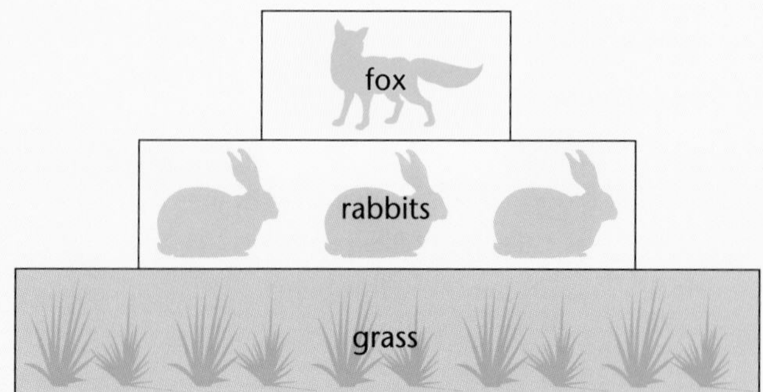

Are you ready for the next step?

a A lion eats a zebra which eats grass. Represent this information in a food chain.

b Which member of this food chain is:
(i) a producer?
(ii) a primary consumer?
(iii) a secondary consumer?

c Which member of this food chain is most likely to be:
(i) a herbivore?
(ii) a carnivore?

d Look at these food chains. Draw them together as a food web.

grass → cows → humans
cereal plants → rabbits → humans
grass → rabbits → humans
cereal plants → pheasants → humans
cereal plants → chickens → humans
water plants → fish → humans
water plants → ducks → humans

e Draw a pyramid of numbers for the food chain below. You can see the numbers of each species in brackets.

barley →	field mice →	owl
(30)	(3)	(1)

f Look at this tree. Suggest why all the energy is not passed on from the tree to the herbivores.

Do you really understand?

Farmers work with food chains that include humans. If they can reduce the amount of energy that leaves between the stages of the food chain, and if they can increase the amount of energy entering the food chain in photosynthesis, they can make more food for humans, and make more money. But sometimes making a change in one food chain can affect organisms in other parts of the food web.

1 Farmers often use weedkillers, chemicals that kill weeds. Weeds often grow taller than cereal plants like barley. Using what you know about photosynthesis, explain why using weedkillers can increase the yield a farmer gets from his crop.

2 Look at this food web.

a If you were a farmer, which animals would you want to remove to increase the number of wheat plants that are eventually eaten by humans?

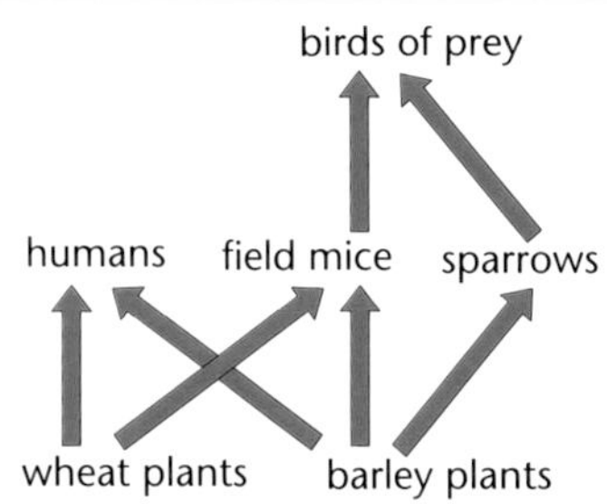

b What effect would this have on the number of birds of prey, and on the number of sparrows?

c A farmer does not like killing small mammals, and plants more wheat plants instead. This way he thinks he'll still get the crop yield he needs, even if some of the plants are eaten. Why is this not a long-term solution?

3 Maeve owns a piece of woodland. Every summer she sprays her trees with pesticide to kill the insects that eat the leaves. The pesticide kills most of these insects but a few survive, carrying small amounts of toxin.

a Look at this food chain.

insect → thrush → cat

Explain, using ideas about toxin build-up, why pesticides could eventually kill Maeve's cat if she keeps spraying for long enough.

b Insects also eat the leaves after they have fallen off the trees in the autumn. They break the leaves down into small pieces. Normally the small pieces are broken down even more by bacteria and fungi until they become completely mixed up with the soil.

(i) How easy will it be for the leaves to break down into the soil after Maeve's use of pesticide?

(ii) Explain why the trees start to show nutrient deficiency symptoms the next spring.

Booster 3

Physical changes

Focus on:

- Particles 1

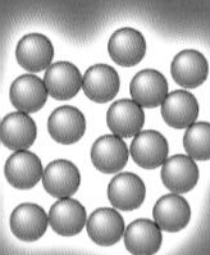

Particles 1

Focus on particles 1

Do you know the basics?

Made from particles

Everything is made from tiny **particles**. The three **states of matter** – **solids**, **liquids** and **gases** – behave in different ways because their particles are arranged differently.

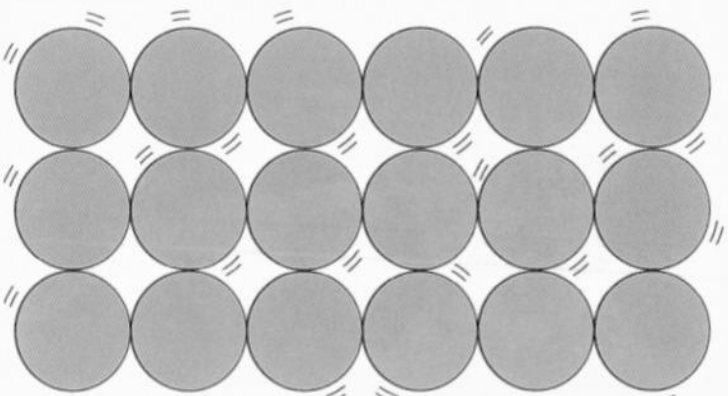

In solids the particles:

- are in rows
- are touching
- vibrate on the spot
- are held together.

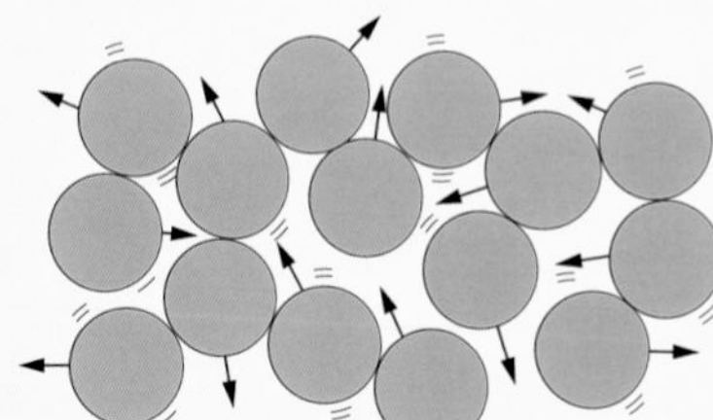

In liquids the particles:

- are disordered
- are touching
- vibrate and slide over each other
- are held together.

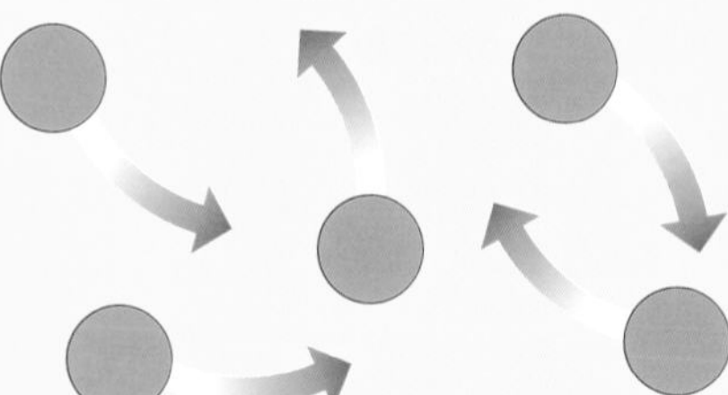

In gases the particles:

- are disordered
- are far apart
- zoom about
- are not held together.

The particle model can be used to explain many events.

- **Diffusion** happens in gases and liquids when particles spread out and mix with other particles.
- The gas particles inside this balloon are moving in all directions and constantly hitting the rubber. Every time a particle hits the rubber, it gives it a tiny push. The sum of all these forces on the area of the balloon is called **gas pressure**.

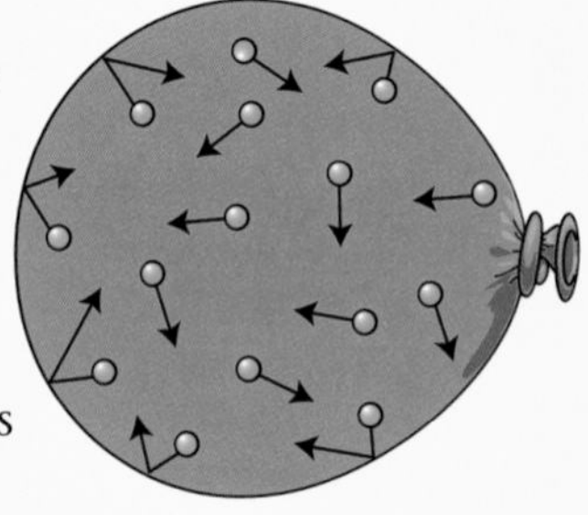

- An increase in temperature makes the particles in matter move more. They take up more space, so the matter **expands**. A decrease in temperature makes the particles move less. They take up less space, so the matter **contracts**. *Remember*: the particles themselves do not change size.

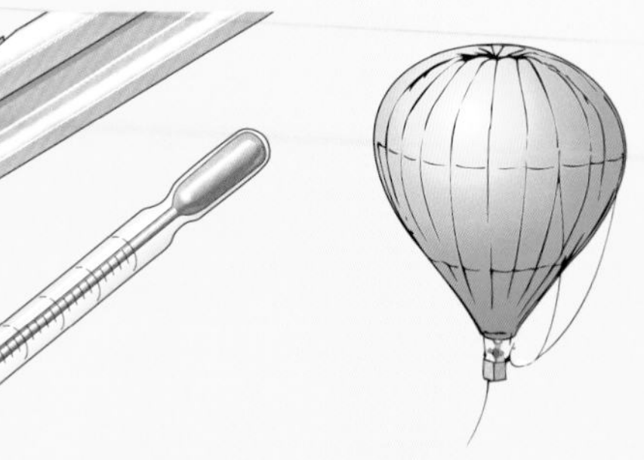

Change of state

When a **physical change** happens no new substances are made. The change is also **reversible**. Changes of state are physical changes. As a solid is warmed up, the thermal (heat) energy makes the particles vibrate faster. The temperature rises.

Once the temperature reaches the **melting point**, all the particles are vibrating so much that they overcome the **forces of attraction** between them. When the temperature reaches **boiling point**, all the particles are moving so much that they break the forces of attraction between them. They move away in all directions as a gas.

Do you remember?

Evaporation happens at the surface of a liquid where only some of the particles gain enough energy to change into a gas. Evaporation takes place at any temperature, not just at the boiling point.

If the gas is cooled, the particles lose energy and move more slowly. Some of them get close enough for forces of attraction to form between them. This is called **condensation**. If a liquid is cooled, the particles lose energy and move more slowly. Stronger forces of attraction form between the particles and they make the regular arrangement of a solid. This is **freezing**.

Rachel and Gemma took some crushed ice out of the freezer. They put it in a beaker and left it in a warm room. This graph shows how the temperature inside the beaker changed.

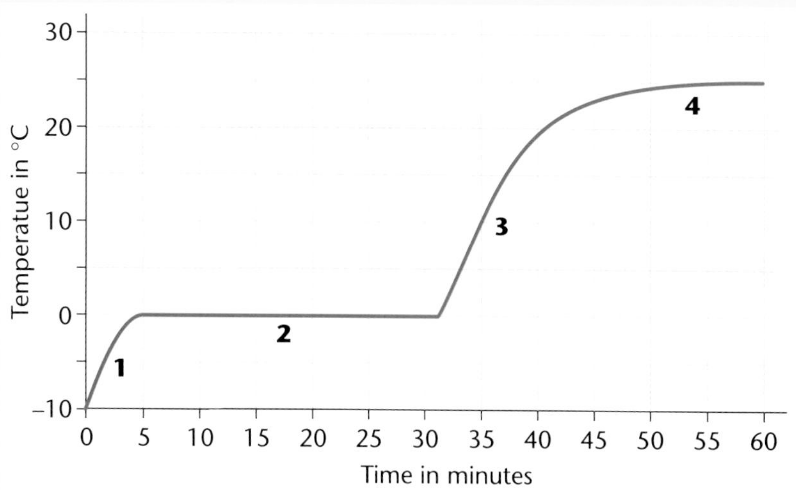

Are you ready for the next step?

a **Draw labelled diagrams to show how the particles are arranged in:**
(i) a solid
(ii) a liquid
(iii) a gas.

b **Look at the graph above and describe how the particles are moving at 1, 2, 3 and 4.**

c **What is happening to the ice at 2?**

d **Why do you think the temperature stays the same at 4?**

Do you remember?

If you give thermal energy to the particles they move faster.

Do you really understand?

1 Describe what happens to the water particles when:

a ice melts

b water boils

c steam condenses.

2 **a** Use the idea of diffusion to explain why the smell of perfume gets weaker as you move further away from where it has been sprayed.

b Use the idea of gas pressure to explain why a balloon keeps its shape.

c Explain why power cables seem to have more slack on a hot day than a cold day.

3 Jane pumps up a bicycle tyre. The tyre pressure rises as she pumps air into the tyre. Give one reason, in terms of the motion of the gas molecules in air, why the pressure rises.

4 **a** The melting points and boiling points of four elements are shown in this table. Copy and complete the table to give the physical state – solid, liquid or gas – of each element at room temperature, 21 °C.

Element	Melting point in °C	Boiling point in °C	Physical state at room temperature, 21 °C
bromine	−7	59	
chlorine	−101	−34	
fluorine	−220	−188	
iodine	114	184	

b Bromine can be a **solid**, a **liquid** or a **gas** depending on the temperature. In which physical state will 10 g of bromine store the most thermal (heat) energy?

c Is bromine a solid, a liquid or a gas when the arrangement of particles is:

(i) far apart and random? **(ii)** close together but random? **(iii)** close together in a regular pattern?

Chemical changes

Focus on:
- Particles 2

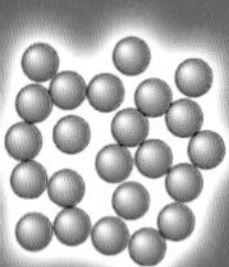

Do you know the basics?

What are substances made of?

Atoms are the simplest type of particle that exists. Atoms cannot be broken down any smaller.

A **mixture** contains more than one substance physically mixed together. Examples of mixtures are air and sea water. Mixtures can be separated by filtration, chromatography and distillation.

Molecules are groups of two or more atoms chemically combined together. They can be made up of just one type of atom or more than one type.

A **compound** is a substance that is made up of more than one type of atom chemically joined together. Examples of compounds include water, iron oxide and sodium chloride. The elements in a compound can be rearranged by chemical reactions

An **element** is made up of only one type of atom. Examples of metallic elements include copper, iron, mercury and zinc. Oxygen, hydrogen and sulphur are non-metallic elements.

When you burn magnesium it reacts with oxygen in the air to make magnesium oxide. This is a **chemical change**. During a chemical change you make new substances. The magnesium has gone. It has changed into magnesium oxide. It is an **irreversible** change. You cannot turn magnesium oxide back into magnesium. Chemical changes happen when the atoms in the **reactants** are rearranged to form new substances, the **products**.

magnesium + oxygen → magnesium oxide

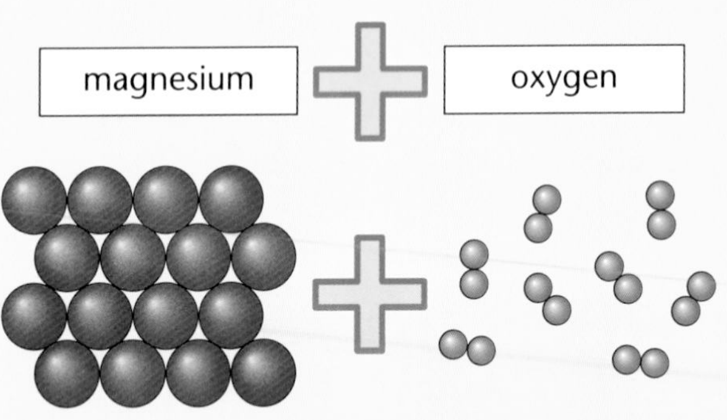

Chemical reactions

There are some types of chemical reaction you need to know about.

Displacement
A more reactive metal can push a less reactive metal out of a compound.

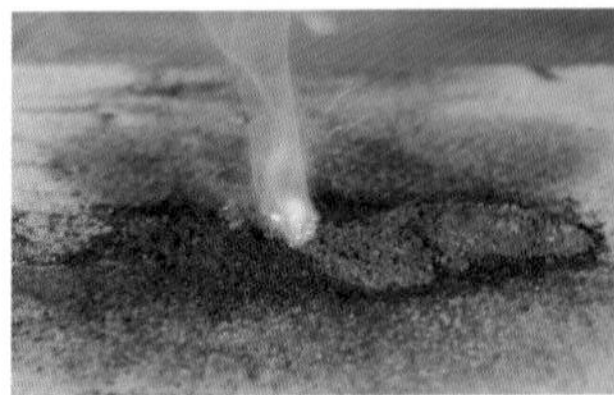

zinc + copper oxide → zinc oxide + copper

Combustion
This is a reaction between a fuel and oxygen.

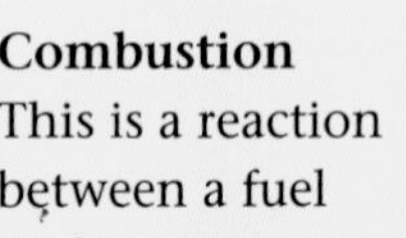

methane + oxygen → carbon dioxide + water

Oxidation
Oxygen may react with metallic or non-metallic elements to produce an oxide.

carbon + oxygen → carbon dioxide

Neutralisation

Alkalis are bases that dissolve in water. They react with acids to make a salt and water.

sodium hydroxide + sulphuric acid →
sodium sulphate + water

Patterns in chemical reactions

There are patterns in some chemical reactions that form compounds. We can write general word equations for them.

A metal + oxygen → oxide
B metal + sulphur → sulphide
C metal + acid → salt + hydrogen
D carbonate + acid → salt + carbon dioxide + water
E acid + base → salt + water
F oxide + acid → salt + water

Are you ready for the next step?

a **Which of these substances are elements and which are compounds?**

water iron sugar
oxygen hydrogen
carbon dioxide

b **Air is a gas at room temperature. The chemical formulae below show some of the substances in air.**

Ar CO_2 H_2O N_2 Ne O_2

Copy and complete this table. Put the formulae in the correct columns to show which substances are elements and which are compounds.

Element	Compound

Do you really understand?

Particles in chemicals

The diagrams below represent the way atoms are arranged in six chemical substances.

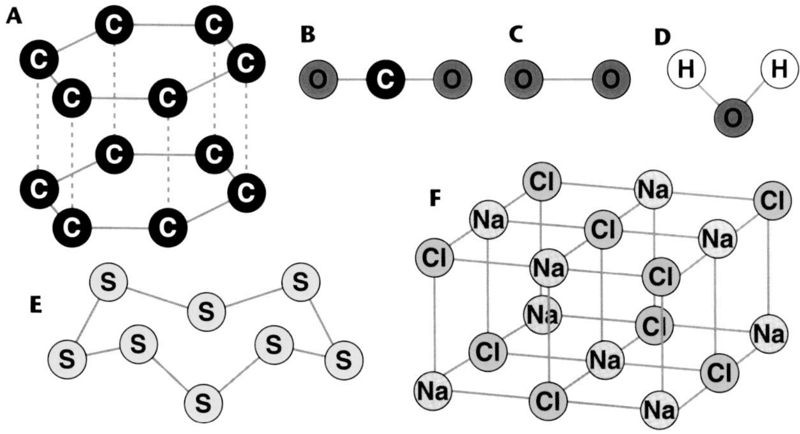

1 **a** Write down the letters that represent the structure of chemical elements.
b Explain how you made your decision.

2 **a** Write down the chemical formulae of two of the compounds represented by the diagrams.
b Is it possible to separate the elements in a compound using their physical properties? Explain your answer.
c How many different sorts of particle are there in substance F?
d How would you separate a mixture of F and D?

3 Air is a mixture of gases.
a Write down the names of an element and a compound that you would expect to find in the air.
b Draw diagrams of how the particles are arranged in the element and the compound you named in **a**.

4 Use the general word equations above to help you with this question.
a Write a word equation for the oxidation of aluminium.
b Name the products formed from the combustion of ethanol.
c Copy and complete this word equation to show the reactants in this displacement reaction.

_________ + _________ → copper + iron sulphate

d Copy and complete the symbol equation below to show the oxidation of sulphur to sulphur dioxide.

$S + O_2 \rightarrow$ _________

Booster 5

Forces all around

Focus on:

- Forces

Do you know the basics?

We measure forces in **newtons**, **N**, and we show them as arrows. The length of the arrow shows the size of the force and the arrowhead shows its direction. Forces act in pairs and work in opposite directions.

During snowmobiling, **thrust** pushes the snowmobile forward and **friction** acts in the opposite direction. If these forces are equal and opposite, or **balanced**, the snowmobile travels at a steady speed. If they are unequal, or **unbalanced**, the snowmobile speeds up, slows down or changes direction. In the drawing, the thrust is 800 N and the friction is 900 N. This means that the **resultant force** is 900 – 800 = 100 N in the backwards direction. The snowmobile slows down.

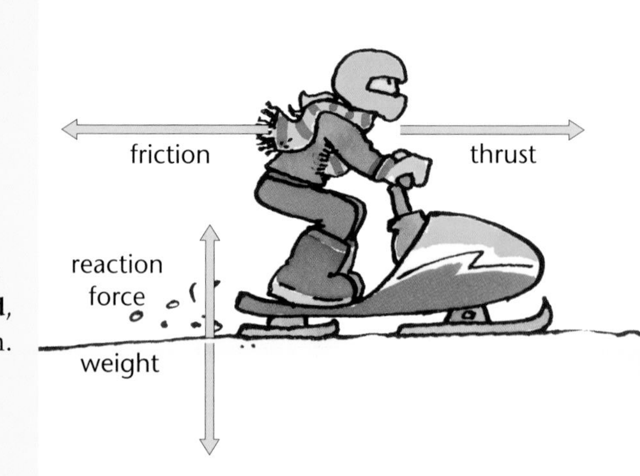

There is another pair of forces on the snowmobile, **weight** and **reaction force**. These forces are balanced, so the snowmobile does not sink into the snow. The **pressure** the snowmobile exerts depends on its weight and the area in contact with the snow.

Freda steers the snowmobile using the handlebars. You turn the front of the snowmobile by pushing or pulling the handles. This produces a **turning force** or **moment**. The longer the handlebar, the larger the turning force. Freda keeps the snowmobile going in a straight line by balancing the clockwise and anticlockwise moments.

Do you remember?

$$\text{speed in m/s} = \frac{\text{distance in m}}{\text{time in s}}$$

(Speed can also be measured in km/h or miles/hour.)

$$\text{pressure in N/m}^2 = \frac{\text{force in N}}{\text{area in m}^2}$$

moment (turning force) = force × distance from pivot

Are you ready for the next step?

a **The snowmobile is moving at a steady speed. The thrust is 800 N. What is the friction force?**

b **The snowmobile travels 20 metres in 5 seconds. What is the speed of the snowmobile?**

c **The snowmobile continues at the same speed for another 10 seconds. How far does it travel?**

d **The snowmobile then covers the next 20 metres at 5 m/s. How long does that take?**

e **Freda hits a patch of poor snow. She will slow down, but she increases the thrust to 1050 N to get through the poor snow. The friction is now 960 N.**
- **(i) What is the new resultant force on the snowmobile?**
- **(ii) What will happen to Freda's speed?**

f **Disaster! Freda loses control and ends up on ice over a lake. The engine stalls. When she restarts the engine, the tracks spin around but she does not go forward. Explain why the snowmobile does not move forward on the ice.**

g **What force balances the weight of the snowmobile, stopping it breaking through the ice?**

h **Which exerts more pressure on the ice, Freda and the snowmobile together or Freda standing on the ice? Use the information in the table to calculate both pressures to work out your answer.**

Freda's weight	800 N
Snowmobile weight	1600 N
Area of Freda's feet	0.12 m²
Area of snowmobile in contact with the ice	0.64 m²

i **Use your knowledge of pressure to suggest and explain how Freda should get back to the bank.**

Do you really understand?

Freda is standing waiting for her turn in the pole vault.

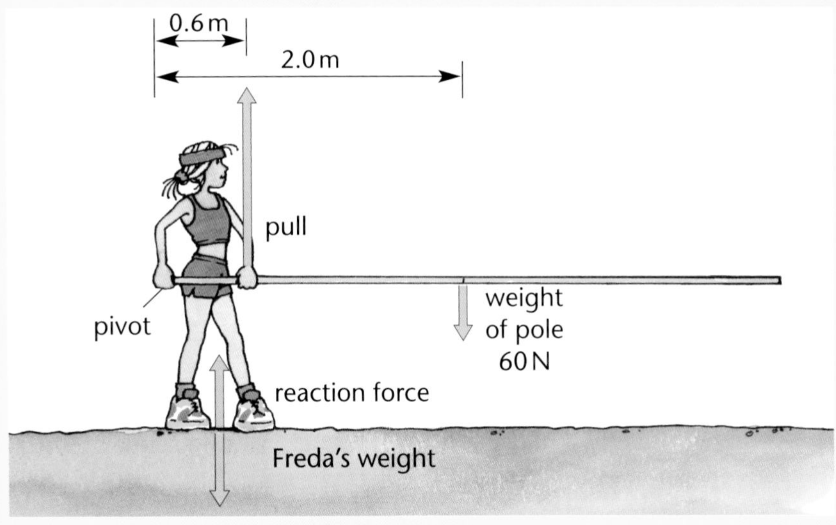

1. What is the size of the reaction force that balances Freda's weight?
2. Use your knowledge of moments to calculate the size of the force Freda is exerting on the pole to balance the weight.
3. It is time for Freda's vault. Her run-up is 12.4 m long and she sprints that distance in 3.7 s. What is Freda's average speed?
4. Does she travel at this speed for the whole 12.4 metres? Explain your answer.
5. She then plants her vaulting pole. Her kinetic energy is transferred into a store of strain energy in the pole: the pole bends. What force causes the pole to bend?

The pole then springs back, throwing Freda into the air.

6. Use the combined weight of the pole and Freda (Freda weighs 800 N) to work out the minimum force needed to accelerate Freda upwards.
7. Freda is catapulted into the air. Once the pole has straightened, Freda's movement upwards will slow. What downwards force causes this slowing?
8. Freda reaches the top and flips herself over the bar. She then falls down the other side. What force causes Freda to fall down the other side?
9. Freda falls into the mat on her back. She then stands up. Explain why Freda sinks further into the mat when she stands up than when she is on her back.
10. On the way back to the team bus Freda carries one of her poles with her kit bag hanging on one end, as shown in this diagram. What is the weight of Freda's kit bag?

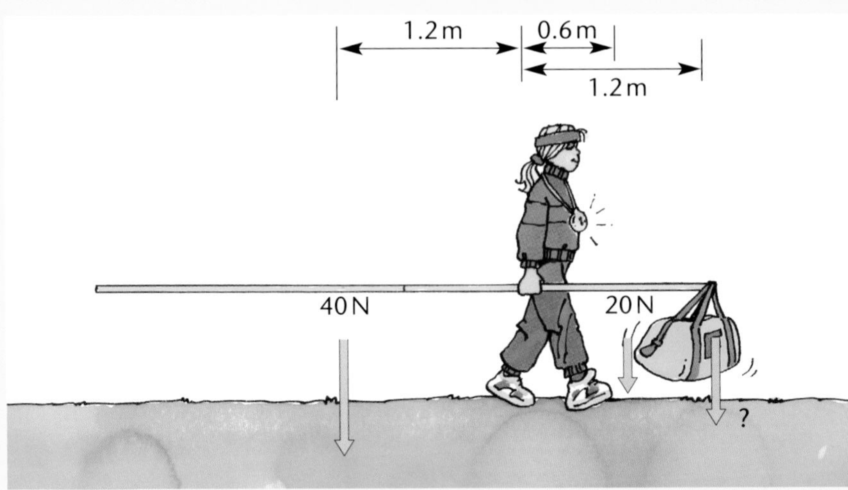

Making things happen

Focus on:

- Energy

Do you know the basics?

Energy makes things happen. Energy can be **transferred** or **stored**.
We measure energy in **joules, J**, or **kilojoules, kJ.** 1000 J = 1 kJ.

The lamp takes in electrical energy and gives out thermal energy and light energy.

The Bunsen burner uses a store of chemical energy and gives out thermal energy and light energy.

The Walkman transfers energy from a store of chemical energy to electrical energy and then to sound energy.

The wind turbine is taking in kinetic energy and produces electrical energy.

a **Explain how:**

(i) you make the lamp brighter
(ii) you heat the water quicker
(iii) batteries 'run down'
(iv) a wind turbine/generator works.

Use the words 'energy' and 'transfer' in each explanation.

Are you ready for the next step?

Energy is **conserved**. This means that it is not created or destroyed, just transferred somewhere or stored. This means that if we put 100 J in, we will get 100 J out.

Energy is often **dissipated** during energy transfers. This means that it ends up spread about, usually as thermal energy. Energy that has been dissipated is difficult to use, because it is spread out between many millions of particles.

During any energy transfer, the percentage of energy that ends up where we want it (not dissipated) is the **energy efficiency**.

Think about the electric lamp shown in the cartoon above. For every 60 J per second entering the lamp as electrical energy, only 5 J end up as light energy. The other 55 J are dissipated as thermal energy. The energy efficiency of the lamp is:

$$\frac{5}{60} \times 100 = 8\%$$

8% is a very low energy efficiency.

Modern energy-efficient lamps are more energy efficient. For example, a 20 W halogen lamp (20 J transferred per second) gives out more light energy than an ordinary 150 W filament lamp.

b **A halogen lamp takes in 20 J of electrical energy every second. It gives out 16 J of light energy.**

(i) How much energy is wasted? Explain how you worked out the answer, using the word 'conserved' in your explanation.
(ii) Where does the wasted energy end up? Use the word 'dissipated' in your answer.
(iii) What is the energy efficiency of the halogen lamp? Show your calculation.

Fossil fuels are non-renewable. This means that they will run out. At the present rate that we are using them, we will run out of oil in about 2030, natural gas in about 2050 and coal in about 2230.

The less energy we waste, the longer the fossil fuels will last.

Imagine you are asked: 'Explain how replacing all the filament lamps in your home with halogen lamps would make fossil fuels last longer.' This is not an easy explanation to write. The explanation has many steps.

a Use the information you have been given.

Halogen lamps are 75% energy efficient. Filament lamps are only 8% efficient.

b Explain what this information means. Use scientific words correctly in your explanation.

This means out of every 100 J of electrical energy put into a halogen lamp, 75 J end up as light energy and only 25 J are wasted when they are dissipated as thermal energy. For a filament lamp, only 8 J out of every 100 J put in end up as light energy and the other 92 J are dissipated as thermal energy.

c Make the first step in linking up the explanation.

So to get the same amount of light energy you need less electrical energy if you use halogen lamps.

d Add in important facts using your scientific knowledge and understanding.

Most electricity is produced in power stations that burn fossil fuels.

e Make the last step.

This means that less fossil fuels have to be burnt in the power station, because less electrical energy is needed.

It is very important to move step by step through the explanation. It is also very important to use scientific words.

Do you really understand?

Gaynor is buying a new washing machine. Model A washing machine has an energy efficiency of 27%. Model B washing machine has an energy efficiency of 32%.

1 Which model of washing machine should Gaynor buy if she cares about making fossil fuels last longer?

2 Explain how buying this model will make fossil fuels last longer. (Follow the five steps suggested above to write a full explanation.)

Using the more energy-efficient washing machine will save Gaynor £15 a year through reduced energy costs. However, model B costs £60 more than model A.

3 For how many years will Gaynor have to use the washing machine to 'pay back' the difference in price?

4 If the washing machine lasts 10 years, how much money will Gaynor save?

How to revise

Key Stage 3 tests

At the end of Year 9 you will do a Key Stage 3 test which covers everything you have studied over the last three years. It is an opportunity for you to show how much you have learned over the course. To help you get a good mark and feel confident about doing the test, it is a good idea to revise thoroughly before the test. Your teacher will probably help you with this in your science lessons, but there is a lot you can do yourself.

Where to revise

- It is best to revise in a room with no distractions like a TV, music or people busy doing other things.
- Most people find it best to have a quiet place for revising.
- Use a table or desk which gives you plenty of space to lay out your books and notes.
- Make sure you have a good source of light to read by.
- Get yourself organised – have plenty of blank paper and a selection of pens and pencils in different colours as well as the notes or books you need.

When to revise

- Try to set aside some time early each evening. Don't leave it too late so that your brain is tired.
- Revise for about 15 minutes and then take a 5-minute break. You could perhaps allow yourself to listen to a song (only one!). Then do another 15 minutes' revision and have another short break. Revise for another 15 minutes, and then have a longer break.
- Breaking up your revision into small chunks like this is much better than revising for a solid hour without any breaks. You will remember more this way.
- Keep a clock close by to help you keep track of the time.

What to revise

You may have already worked through the six double-page booster spreads to help you fully understand the five key ideas in science (cells, interdependence, particles, forces and energy). Even if you have, we suggest you work through them again to improve your knowledge of these key ideas so you can explain other topics.

The four revision spreads have SAT questions with notes from an examiner giving you suggestions about how to help you focus your revision on answering SAT questions.

You need to understand ...	... to understand and explain ...
cells	reproduction, photosynthesis and digestion
interdependence	food chains, food webs and energy transfer
particles	the properties of solids, liquids and gases, diffusion and pressure in gases and liquids
particles	what atoms and molecules are made of, how substances react to make new substances
forces	why things move, change direction or speed or balance, and how planets orbit the Sun
energy	how energy is transferred to make things happen, how electrical devices work and how things heat up and cool down.

Revision timetables

- Don't try to revise your entire science course in one night!
- Plan your revision long before your test.
- Work out how you will divide the material up, and how much you will revise each night.
- Work out how many evenings you will have available for revision.
- Make a timetable something like this to make sure you cover every topic at least once.

Day	What I will revise	Tick when done
Day 1	Topic 1	
Day 2	Topic 2	
Day 3	Topic 3	

How to revise

There are many different techniques that you can use. Here are just a few.

1 Read – Cover – Write – Check
Read an entire double page spread in the book. Then close the book and write notes on as many key points as you can remember. Then open the book again, check what you wrote down and go over the things you didn't remember. Repeat this until you can remember everything on the pages.

2 Make a memory map for each section. Then try to learn the memory map – think about the way each part of it is linked together. Then cover the map up and try to redraw it.

3 Write out lots of questions. Then close the book and see if you can answer them. You can also get someone else to ask you the questions. On the next few pages there are example test questions with tips from the examiners about the best way to answer.

4 Make sure you know the meanings of all the key words you come across.

5 Make up silly rhymes or mnemonics for important facts or patterns. The sillier they are, the easier it will be for your brain to remember them.

Don't just sit there with a book in front of you. It's not the best way to learn. The best way to revise is by actively doing tasks to make your brain work. This will make it much easier to remember things. Then you can go into your test confident that you will do the best you can.

Scientific enquiry example questions

The example questions show you how to write good answers to make sure you always get all the marks available.

The main reasons why pupils do not do as well as they should is that they give answers that are too general, or that are incomplete and do not give a full answer to the question.

Do not fall into this trap. Read these extra comments around the questions for useful tips that will help you to get all the marks and make sure you are successful in your KS3 Test.

13 A Belgian scientist, van Helmont, did a famous experiment in the seventeenth century. He planted a young willow tree in a pot of dry soil.

During the next five years he added nothing except rainwater.

He then removed the tree from the pot and weighed the tree. He dried and weighed the soil.

When weighed	Mass of willow tree in kg	Mass of dried soil in kg
at the start	2.3	90.6
after 5 years	76.7	90.5

Van Helmont's conclusion was that the increase in mass of the tree had only come from the rainwater.

a Why did van Helmont weigh the soil at the beginning and the end of the experiment? (1)

So that he could measure any change in the mass of the soil ✓

b Why did he dry the soil carefully before he weighed it? (1)

Any water present would have affected the mass ✓

c What **two** pieces of evidence did he use to reach his conclusion? (2)

Only water was added ✓

The mass of the soil only went down slightly ✓

Van Helmont put a lid on the pot to keep the dust out. He punched holes in the lid to let the rainwater in.

d Why was it important to keep dust out of the pot? (1)

Dust would have increased the mass of the soil ✓

Total (5)

a Good experimental design means thinking about how to make your results reliable. Van Helmont needed to be sure that the increase in mass of the tree had not come from the soil.

b The amount of water in soil can vary and this would change the mass of the soil.

c Make sure you give **both** answers. Conclusions must be based on the evidence from an experiment, so being able to see what the evidence is telling you is important.

d Over 5 years the amount of dust would have changed the mass of the soil. But van Helmont got his design wrong – the dust could still get through the holes!

14 Clare melts some stearic acid in a beaker. She uses a water bath to do this.

a Why is it better to use a water bath to melt the stearic acid? (1)

The stearic acid might be inflammable ✓

> a Safety is very important, so think about how an experiment can be made safer when you plan it.

Clare takes the beaker out of the water bath and uses a temperature probe to record the temperature of the stearic acid as it cools.

b Give **two** reasons why it is better to use a temperature probe rather than a thermometer in the experiment. (2)

The temperature probe makes continuous readings ✓

The graph is drawn automatically ✓

> b Make sure you give both, and that each one is explained. Writing 'more accurate' will not get credit. A probe is more precise than a thermometer, so this is a third reason. There is less chance of anomalous results, a fourth reason.

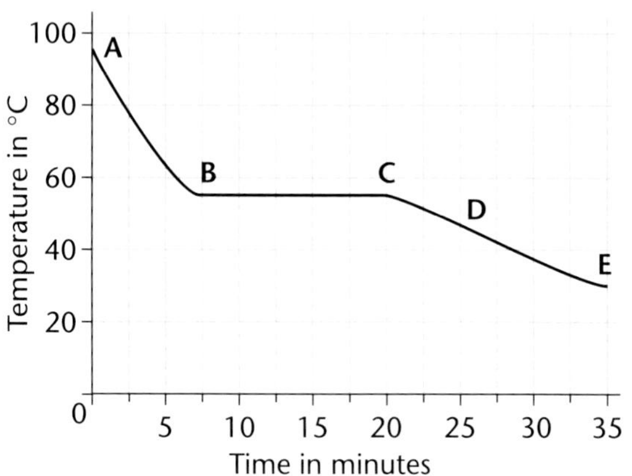

The graph shows how the state of the stearic acid changes as it cools.

c What is the state of the stearic acid at point D on the graph? (1)

Solid ✓

> c Being able to look for patterns and interpret results is an important skill, so this kind of question tests that skill.

d Which letter on the graph shows when the stearic acid **starts** to change state? (1)

B ✓

Total (5)

> d This is another way of seeing if you can interpret results. At A the stearic acid is still liquid; at C it is completely solid. This graph is a **cooling curve**.

Biology example questions

2 The diagram shows a plant cell.

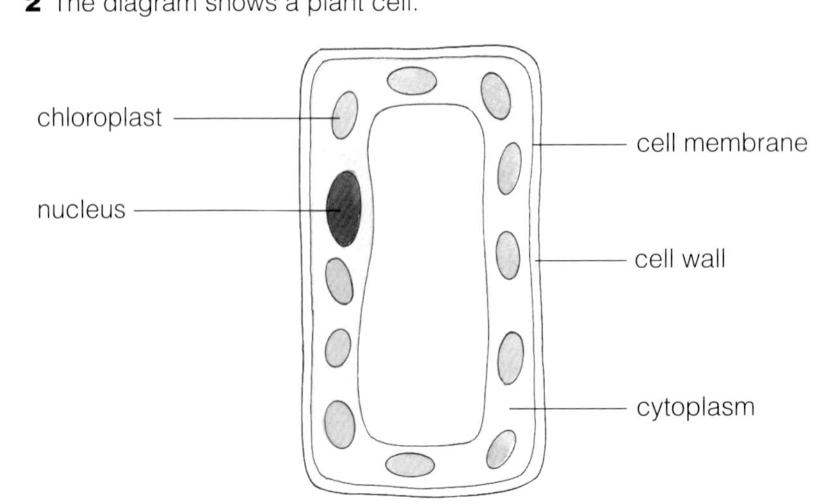

a Where in the plant would you find this type of cell? (1)

In the leaf ✓

b Most animal and plant cells have a nucleus. Give two other parts, labelled on the diagram, which are present in both animal and plant cells. (2)

Part 1 Cell membrane ✓

Part 2 Cytoplasm ✓

c i What is the function of the cell wall? (1)

To give shape to the cell ✓

ii What is the function of the chloroplasts? (1)

Photosynthesis occurs in the chloroplasts. Carbon dioxide and water are converted into glucose and oxygen ✓

Total (5)

a The cell is called a palisade cell. It makes up the tissue under the upper surface of the leaf. This is one of several types of cell you need to know about and to recognise in diagrams.

b Remember, plant cells have chloroplasts, a cell wall and a large vacuole, which animal cells don't.

c i The cell wall is made of cellulose. It provides support for the plant cell. Cell walls do not hold the cell contents together – this is the function of the cell membrane. Cell walls allow plant cells to stack together to make the plant.

c ii 'Photosynthesis' is the word that will get you the mark. Do not write 'to make food for the plant', as this is too vague.

Light is absorbed by the chlorophyll in the chloroplasts. This provides the energy for the photosynthesis reaction.

4 Ben copied the following information from the labels of two packets of food.

Food	Energy in kJ/100 g	Protein in g/100 g	Fat in g/100 g	Sugar in g/100 g	Fibre in g/100 g	Vitamin C in mg/100 g
X	424	6.9	0.6	3.6	6.2	0
Y	736	20.2	10.6	0	0	0

a Food Y contains a smaller variety of nutrients than food X. Give **two** reasons why food Y might be chosen instead of food X as part of a balanced diet. (2)

Reason 1

The rest of the person's diet might be low in protein ✓

Reason 2

The person could be underweight or ill and need a high energy intake to build them up ✓

b Ben made a curry. The label on the curry power showed:

	Calcium in mg/100 g	Iron in mg/100 g	Vitamin C in mg/100 g
curry powder	640	58.3	0

i Give one reason why calcium is needed by the body. (1)

For healthy bones ✓

ii Give one reason why iron is needed by the body. (1)

To make red blood cells ✓

c With his curry, Ben also had:
boiled rice; chopped-up, boiled egg; a glass of water; a slice of lemon.
Which one of these foods provided Ben with vitamin C? (1)

Lemon ✓

Total (5)

a The question 'Why choose food Y?' requires reasons or explanation to be given.

You need to make comparisons between food X and food Y in order to give reasons why you would choose Y. Just stating 'Food Y is much higher in energy, protein and fat' is a poor answer.

b i 'Strong *or* healthy bones *or* teeth' is a good, complete answer.

Writing the word 'bones' or 'teeth' would probably get you the mark, but the question asks for a reason.

b ii Iron is needed to make in red blood cells. If you do not have enough iron you become anaemic.

Do not write just 'blood' or 'blood cells' – this does not give enough detail to get you the mark.

c Vitamin C comes from fruit, particularly citrus fruit such as lemons and oranges.

Revision 4

Chemistry example questions

a These terms are ones you need to remember.

Remember that water melts and freezes at its melting point.

Water boils and condenses at its boiling point.

3 Water can exist in three physical states: ice, water and steam.

a What is the name given to the process that changes: (2)

i ice into water? . melting ✓

ii steam into water? condensing ✓

b A beaker of ice was placed in a warm room. The graph shows how the temperature in the beaker changed from the start of the experiment.

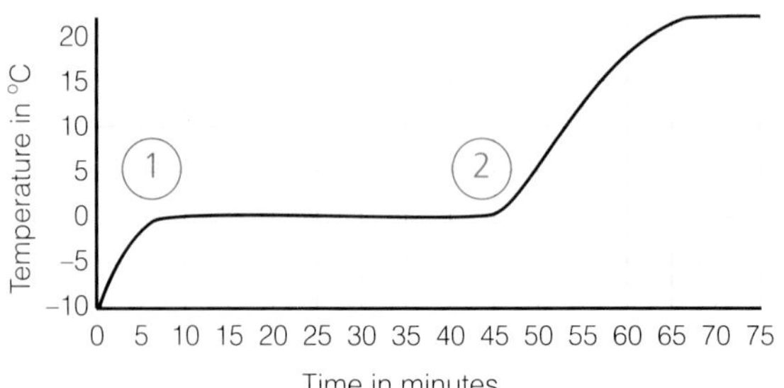

How long after the start had all the ice just gone? (1)

45 minutes

b Look at the graph.

1 The ice starts to melt after about 6 minutes.

2 The temperature does not start to go up until all the ice has melted at 0 °C.

c Draw diagrams, in the boxes below, to show the arrangement of water molecules in ice, water and steam. Use circles, like this ○, to represent the water molecules.

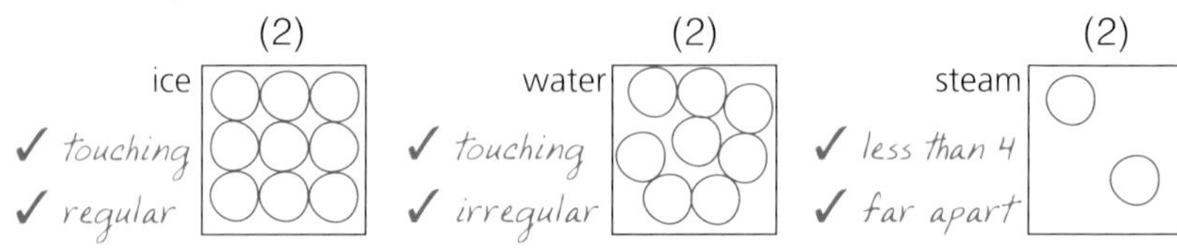

d Water is a compound of two elements. A diagram of a water molecule is shown below. In this diagram the circles represent atoms.

In the circles, write the correct symbols for the elements. (1)

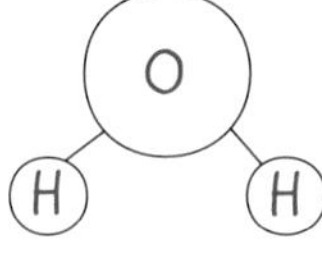

Total (10)

c You will get two marks for each drawing. The ticks show what you will get the marks for.

Remember to draw circles at the size indicated. Try to draw them roughly the same size. You will lose a mark if they are of very different sizes.

The particles in solids are regularly arranged and close packed. The circles can alternate rather than lining up, e.g.

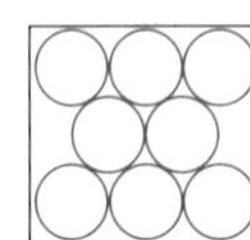

The particles in liquids are random or irregular, but close packed. You cannot squeeze a liquid into a smaller volume. In your drawings make sure there are not too many spaces and that each circle touches at least two other circles.

Gases are mainly empty space with particles racing about and bouncing off each other. Show no more than three circles in your drawing.

d The formula for water is H_2O. If you think about the number of atoms here, the two smaller circles must be Hs for hydrogen and the larger one must be O for oxygen. You are not expected to remember this shape but you ought to be able to think out the answer.

5 Alex has four solids. They are labelled **W, X, Y** and **Z.** He adds a sample of each one to some dilute acid. The table shows his results.

Solid	Result with dilute acid
W	it reacts slowly and gives off hydrogen
X	it reacts quickly and gives off carbon dioxide
Y	it dissolves and the liquid becomes warm
Z	it remains undissolved as a white powder

a i Which solid could be chalk (calcium carbonate)? (1)

X ✓

ii Give the name of another rock which reacts with acid in the same way as chalk. (1)

Limestone ✓

b i One of the solids is a metal. State which one and give the reason for your choice. (2)

W because it gives off hydrogen with acid ✓

ii The list below gives five metals.

copper gold potassium sodium zinc

Write them in order of reactivity starting with the most reactive. (1)

Most reactive Potassium Sodium Zinc Copper Gold Least reactive ✓

iii Give the name of another metal which would react with acid in a similar way to zinc. (1)

Iron ✓

c As each of the solids reacts, the acid is used up. Describe a test you can use to show whether or not acid is still present. (2)

Add universal indicator. ✓ It turns from green to red if acid is present. It stays green if the acid has been used up ✓

Total (8)

a i All carbonates react with acid to give carbon dioxide.

a ii Chalk and limestone are sedimentary rocks made of calcium carbonate. Marble is another correct answer. Marble is a metamorphic rock made of calcium carbonate.

b i With acid, reactive metals give off hydrogen. They react to form the metal salt. The more reactive they are, the faster they react.

b ii You need to know a reactivity series, e.g.

potassium
sodium
calcium
magnesium
aluminium
zinc
iron
[hydrogen]
copper
silver
gold

Notice hydrogen is in the series to link it with reactions with acids. A metal above hydrogen will react with acids to give hydrogen gas. A metal below hydrogen will not.

c To get the second mark you need to show the result of the test if the acid is still present, as well as the result if the acid is used up.

Writing 'Use indicator' would not get you a mark unless it was obvious from the colour changes you gave which indicator you meant.

b iii If you know your reactivity series, questions like this are very easy. The metal just above or just below zinc will react in a similar way. Metals above aluminium will react much more quickly, and metals below iron will not react.

Physics example questions

8 Jack is pushing a luggage trolley along level ground at an airport.

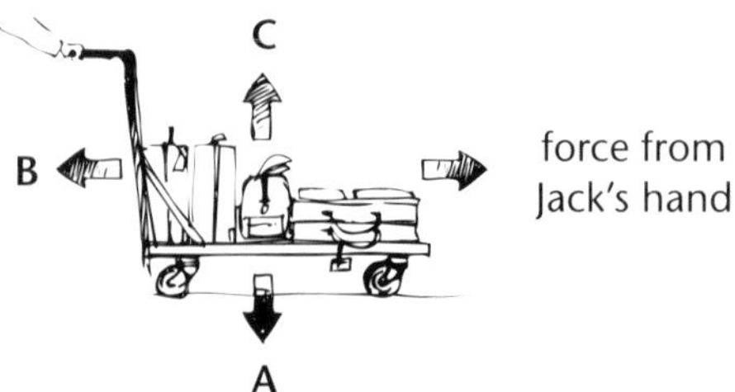

There are four forces acting on the trolley.

a One of the forces is the push from Jack's hands. The others are friction, weight and the reaction of the ground.

Complete the sentences. (3)

Force A is weight ✓

Force B is friction ✓

Force C is reaction of the ground ✓

b What are the units in which force is measured? (1)

newtons ✓

c The trolley is moving forwards, and it is getting faster. One pair of forces is now unbalanced.
Compare the sizes of these two forces. (1)

The forward force of Jack's hand must now be bigger than the backward force of friction ✓

d Jack has to push the trolley 150 m to the check-in desk.
If he pushes the trolley at 3 m/s, how long will it take him? (1)

$\text{Speed} = \frac{\text{distance}}{\text{time}} \quad 3 = \frac{150}{?}$

therefore $? = \frac{150}{3} = 50\,\text{s}$ ✓

Total (6)

a When choosing an answer from a list, always copy correctly and completely. Do not shorten 'reaction of the ground' to 'reaction'.

b The symbol for the unit of force is N (capital). The word is newton (without a capital letter).

c These answers always require a sentence. You need to include the link that to get faster needs a bigger forward force.

If you were unsure about the name of force B, an answer like 'The force from Jack's hand is larger than force B' is acceptable because it uses the information from the diagram.

Do not write 'It is bigger' because the marker does not know what 'it' refers to.

d In any calculation, write down the equation in the form that you remember it and put the numbers in. Then rearrange the equation. Use small numbers to help you check your rearranged equation. For example, $3 = \frac{6}{?}$ is easier to rearrange than an equation with letters involving algebra.

Many speed calculations can be done using common sense rather than the formula:

$$\text{speed} = \frac{\text{distance}}{\text{time}}$$

It always pays to check your answers using common sense. Think to yourself: if it goes 3 m in 1 second then it goes 30 m in 10 seconds or 90 m in 30 seconds. So it goes 150 m in 50 seconds. Do not forget the units!

11 a The Earth is the third planet from the Sun.

i Which is the second planet from the Sun? (1)

Venus ✓

ii Which is the fourth planet from the Sun? (1)

Mars ✓

b The diagram shows the orbits of the Earth and Mercury. Mercury takes 88 Earth-days to orbit the Sun.

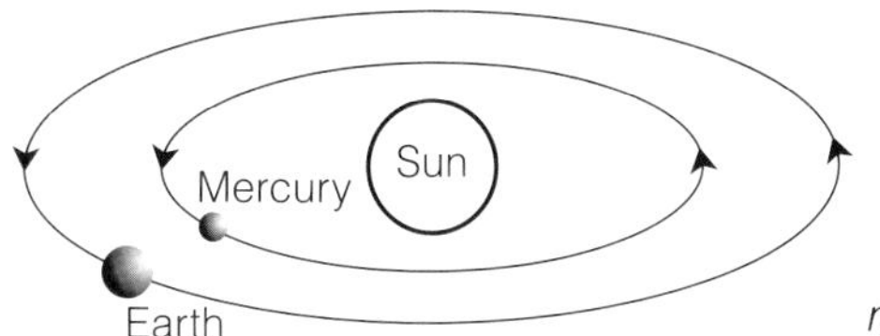

In the diagram, the Earth and Mercury are lined up with the Sun. How long will it take before the Earth and Mercury are lined up with the Sun again? Tick the correct box. (1)

less than 88 Earth-days ☐

exactly 88 Earth-days ☐

more than 88 Earth-days  ✓

exactly 365 Earth-days ☐

c Mercury and Pluto are both small rocky planets. Mercury is one of the brightest objects in the night sky, but Pluto is so faint that it cannot be seen with the naked eye.

Give **two** reasons why Mercury is much brighter than Pluto. (2)

Reason 1

Mercury is nearer the Sun so gets more light than Pluto ✓

Reason 2

Earth is nearer Mercury than Pluto so light has less distance to travel after being reflected ✓

Total (5)

a You need to remember the order of the planets in the Solar System from the Sun, at least to Jupiter: Mercury; Venus; Earth; Mars; Jupiter; Saturn; Uranus; Neptune; Pluto.

b Mercury takes 88 days to get back to the same place (where it is shown on the diagram). In this time the Earth will have moved on about one-quarter of an orbit so Mercury will have to cover at least another quarter of its orbit to catch up with Earth.

c The further light, or any form of energy, has to travel, the more the light or energy spreads out.

Mercury is nearer to the Sun than Pluto. It receives much more light. Light also has less distance to travel from Mercury to Earth than from Pluto to Earth, so spreads out less. Good answers include both parts of the explanation.

Glossary

acceleration Speeding up. Acceleration is shown by an upwards gradient on a speed–time graph.

accuracy How close a measurement is to its 'real' value.

acid rain Rain polluted by acidic gases such as sulphur dioxide dissolved in it. Acid rain is more acidic than rainwater that is not polluted.

addictive Likely to cause addiction – that is, a need to keep taking a drug. Without the drug, an addicted person feels ill.

air pressure The gas pressure caused when air particles all around us hit us and other surfaces.

alloy A metal made of a mixture of metallic elements, or of a metal and a non-metal.

anomaly Something that is more than, less than, or different to what is normal or expected.

antagonistic pair Two muscles that work against each other to move a bone at a joint.

application rate The amount of substance you need to apply to a given area.

artificial insemination Semen is put into the female's vagina through a long tube to make her pregnant without sexual intercourse.

artificial satellite A satellite that is made by people, such as a communications satellite.

asexual reproduction Reproduction that does not involve the fusing of nuclei of two sex cells.

balanced equation A chemical equation where the number of atoms on one side balances the number of atoms on the other side.

bone The hard tissue that makes up the skeleton.

cartilage A smooth substance found on the ends of bones, which allows them to move over each other easily.

cast iron A hard, brittle alloy of iron and carbon that can be shaped by heating and pouring into moulds.

catalytic converter A device fitted to a car's exhaust that removes some of the polluting gases by chemical reactions.

chlorophyll A green substance in plants that is needed for photosynthesis.

clones Organisms that have exactly the same genes. All their genes come from one parent.

compete Try to get the same food sources or other resources as other organisms.

compress To squeeze into less space. When you compress a gas, the particles move closer together and so the gas has a smaller volume.

conduct To pass along or through. Thermal (heat) energy can be conducted. Electricity can be conducted.

conserved Energy – energy is conserved: it is not created or destroyed, but just passes from place to place. We call this 'conservation of energy'.

conserved Mass – stays the same. In all chemical reactions, the mass of substances is conserved.

contract Get shorter. When a muscle gets shorter, we say it is contracting.

counterbalance A weight used to balance another force, that stops something falling over.

cylinder Part of hydraulic and pneumatic machines that acts like a plunger, moving in and out of a cylinder.

deceleration Slowing down. Deceleration is shown by a downwards gradient on a speed–time graph.

deficiency symptoms Noticeable features of poor health or disease caused by a lack of certain nutrients.

dependent variable A variable (something that can be measured) whose value depends on that of something else. Usually the *y*-axis on a graph.

depressants Drugs that slow down the body's reactions and make the user feel drowsy and relaxed. Alcohol is a depressant.

desirable feature A feature that is useful, that you would choose to pass on in selective breeding.

displacement reaction A chemical reaction in which an element is removed from its compound by a more reactive element.

dissipated Spread about. Energy such as light or heat is dissipated from a source to the surroundings.

distance–time graph A graph that shows the speed of a moving object – the distance travelled for each unit of time.

drag The friction force between a moving object and the air or water particles it is moving through.

drug Any substance that changes how your body works, or alters the way you think and feel.

energy efficiency How much energy a device wastes. Something with low energy efficiency wastes lots of energy.

equilibrium When the forces on an object are balanced, it is in equilibrium.

estimate An informed guess, usually applied to numbers.

exert When you exert pressure on something, you apply a force to it.

extinct A species that becomes extinct dies out completely.

fertiliser A substance used to keep soil fertile, so that plants have all the mineral salts they need to grow.

fitness Being in good health, and able to carry out the body's functions efficiently.

generator A device that takes in kinetic (movement) energy and turns it into electrical energy.

genes Instructions that control the way our features develop. Genes are passed on from parents to offspring.

genetic engineering Taking a gene out of one species and putting it into another to give desirable features.

genetically modified (GM) food Food produced from crops or other organisms that have had their genes changed by genetic engineering.

geocentric model A model of the universe with Earth at the centre and everything, including the Sun, moving around it.

geostationary orbit The path around the Earth taken by a satellite travelling at the same speed at which the Earth rotates.

global warming An increase in the average temperature of the Earth.

gradient The slope of a line on a graph.

gravitational potential energy The energy stored because something is lifted up.

haemoglobin The substance in red blood cells that oxygen attaches itself to.

hallucinogens Drugs that cause the user to see things that are not really there. LSD is a hallucinogen.

heliocentric model A model of the Solar System with the Sun at the centre and the planets moving around it.

hydraulic machine A machine that works by transferring pressure through a liquid.

hydrocarbons A substance that contains only carbon and hydrogen atoms.

illegal drugs Drugs that harm the body and are banned by law.

independent variable A variable (something that can be measured) whose value does not depend on anything else. Usually the *x*-axis on a graph.

insecticide A type of pesticide that kills insects.

interact When two or more things have an effect on each other.

joint A place in the skeleton where bones can move.

lactation Producing milk to feed young after they are born.

leaf A plant organ that is important for photosynthesis.

lever A simple machine for lifting objects, that turns around a pivot.

ligament Strong tissue that holds bones together at a joint.

load A force that is carried by a person or machine. The load is often the weight of an object.

magnify To make something appear larger than it is.

mean The average of a set of values or measurements.

medical drugs Drugs given by a doctor or pharmacist to help make someone better.

milk yield The amount of milk a cow produces.

moment The turning effect of a force around a pivot. The moment of a force depends on the size of the force and its distance from the pivot.

monitor To check and measure over a period of time.

muscle Tissue of muscle fibres that can contract (shorten). A muscle is an organ made of muscle tissue and other tissues such as capillaries.

natural satellite A satellite that is made by nature, such as the Moon orbiting the Earth.

newtons per square metre, N/m^2 The unit by which pressure is measured. Also called a pascal.

non-selective Affecting everything. Non-selective insecticides kill all insects, not just the pests.

ore A rock containing a metal or a metal compound.

organic Not involving the use of manufactured chemicals.

oxide A compound formed when a substance burns and joins with oxygen in the air.

pascal Another name for the unit newtons per square metre.

pesticide A chemical used on crops to kill insects or other pests.

pests Destructive insects or other animals that attack crops, food or livestock.

phlogiston A substance supposed by eighteenth century scientists to be contained by anything that burns, and that is released by burning.

photosynthesis Plants make food by photosynthesis. They turn carbon dioxide and water into sugars and oxygen, using light energy.

piston Part of hydraulic and pneumatic machines. Pistons move inside cylinders.

pivot The point around which a lever turns.

pneumatic machine A machine that works by transferring pressure through a compressed gas.

polar orbit The path taken by a satellite passing over the North and South Poles of the Earth.

pollution The contamination of water, air or soil by harmful or toxic substances.

potential difference Another term for voltage. It is the difference in potential energy between two points on an electric circuit.

power rating How many watts (joules per second) of energy an electrical device transfers. For example, a light bulb can have a power rating of 60 watts.

precise Accurate, or as close as possible to an exact amount or detail.

pressure The effect of a force spread out over an area.

principal of moments This states that when two moments are balanced, the sum of the anticlockwise moments equals the sum of the clockwise moments.

probability The chance of an event happening.

raw material The basic material or natural substance from which something is made, such as metal ore.

reactive A reactive substance takes part in chemical reactions, usually quickly and releasing lots of energy.

reactivity series A list of metals arranged in order of reactivity, with the most reactive at the top.

recreational drugs Legal drugs such as caffeine, alcohol and nicotine taken for enjoyment.

red blood cells Special cells that carry oxygen around the bloodstream.

relax When a muscle stops contracting we say it relaxes. It gets longer and thinner.

reliability How much something can be trusted. A value becomes more reliable the more times it is measured.

root A plant organ that takes in water and minerals from the soil, and anchors the plant in the soil.

root hairs Tiny structures on a root that absorb water from the soil.

salt A substance formed in a neutralisation reaction.

Sankey diagram A diagram that shows the amount of energy being transferred. The widths of the lines show the amounts of energy.

satellite An object that orbits a larger object.

selective Affecting some things and not others.

selective breeding Choosing parents with desirable features to produce new varieties of animals or plants that have these desirable features.

self-pollinated When the nucleus of the egg cell in a flower is fertilised by the nucleus from the male part of the same flower.

side effect An additional, often undesirable, effect of a drug on the body of the person who takes it.

skeleton A system of bones that protects and supports your body and allows you to move.

speed–time graph A graph that shows the acceleration, deceleration or steady speed of a moving object – the speed travelled for each unit of time.

sprain Stretch a ligament enough to cause swelling and pain.

steady speed Not getting faster or slowing down, but keeping at the same speed.

steel A hard, strong alloy of iron and carbon that can be bent without breaking.

stimulants Drugs that speed up the body's reactions and make the user feel they have lots of energy. Caffeine is a stimulant.

stomata Holes in a leaf's surface. Gases get in and out of the leaf through stomata.

strain Stretch or pull a muscle enough to cause pain.

streamlined A shape of an object that allows it to move through the air easily. A streamlined car has low air resistance.

synovial fluid A liquid found inside joints that allows the bones to move easily.

synovial membrane A thin lining around certain joints in the skeleton that produces synovial fluid.

tendon Tissue that connects muscles to bones.

thrust The pushing force of a rocket or engine.

toxins Poisonous substances that may cause death or serious injury.

transmit To pass on from one place to another.

turbine A device for changing movement in one direction into spinning movement.

turning effect When there is a force on an object and the force arrow is to one side of the pivot, the force has a turning effect on the object.

unit A standard quantity by which something is measured.

unit of alcohol The amount of alcohol in half a pint of beer, a glass of wine or a measure of spirits.

unreactive An unreactive substance does not take part in chemical reactions, or does so only slowly.

veins In plants, these transport water, minerals and sugars around the plant.

volatile Easily vaporised at normal temperatures.

voltaic cell A simple battery in which electricity passes between two metals placed in dilute acid.

voltmeter An instrument used to measure voltage.

water pressure The pressure in water. It is caused because water pushes on objects from all sides as the water particles collide with the object.

weedkiller A chemical substance applied to weeds, that kills them.

weeds Plants that grow where you don't want them to grow.

withdrawal symptoms The bad feelings that come after the effects of taking a drug have worn off.

Index

Note: page numbers in **bold** are for glossary definitions